CW00349634

PRAISE FOR *SCIENCE IN A DEMOCRATIC SOCIETY*

"Kitcher's focus in this book is on a problem that is both practically urgent and intellectually challenging: how, given the current erosion of trust in scientific expertise, should a commitment to democracy shape the institutions of science and what role should scientific expertise play in democratic political decisions? Kitcher not only performs the invaluable service of framing the issues in a clear-headed way, he also points toward plausible answers to the questions he poses. By his own example, he shows how valuable the role of public intellectual can be."

—Allen Buchanan, James B. Duke Professor of Philosophy
and Professor of Law, Duke University

"Kitcher's new book is an outstanding contribution to the more recent social and political turn of philosophy of science. It's a stunning piece of work and the right kind of study we need."

—Martin Carrier, professor of philosophy,
Bielefeld University, Germany

"Not free inquiry, but the good of society democratically determined is the ideal by which science should be governed, argues Kitcher in this radical yet eminently reasonable book. Essential reading for anyone interested in science, science policy, or the future of the human race."

—Michael Strevens, professor of philosophy, New York University

SCIENCE IN A DEMOCRATIC SOCIETY

SCIENCE IN A DEMOCRATIC SOCIETY

PHILIP KITCHER

59 John Glenn Drive
Amherst, New York 14228-2119

Published 2011 by Prometheus Books

Cover design by Jacqueline Nasso Cooke.

Inquiries should be addressed to
Prometheus Books
59 John Glenn Drive
Amherst, New York 14228–2119
VOICE: 716–691–0133
FAX: 716–691–0137
WWW.PROMETHEUSBOOKS.COM

15 14 13 12 11 5 4 3 2 1

Library of Congress Cataloging-in-Publication Data

Kitcher, Philip, 1947–
Science in a democratic society / by Philip Kitcher.
p. cm.
Includes index.
ISBN 978–1–61614–407–4 (cloth : alk. paper)
ISBN 978–1–61614–408–1 (ebook)
1. Science—Social aspects. 2. Science—Philosophy. 3. Values. I. Title.

Q175.5.K524 2011
303.48'3—dc22

2011014989

Printed in the United States of America on acid-free paper.

For Michael
in friendship

CONTENTS

PREFACE

Since writing *Science, Truth, and Democracy* (Kitcher 2001), I have continued to think about the relations between scientific research and democratic values. During the past decade, I have given lectures and public presentations on related themes, and have published a number of articles. Along the way, I have often thought of developing my ideas more systematically.

An additional impulse to do just that came when, in 2006, I was awarded the Prometheus Prize by the American Philosophical Association. I am extremely grateful to Paul Kurtz, not only for his generosity in endowing the prize, but also for his extraordinary efforts over decades to advance the cause of secular humanism (I very much hope he will feel that this book is in the spirit of his endeavors). I am also greatly honored that the APA Committee on Prizes and Awards chose me to receive it.

In preparing the final version, I have been greatly helped by many people. During the past decade, my ideas have evolved through exchanges and conversations with Nancy Cartwright, Lorraine Daston, Isaac Levi, Helen Longino, and the late Sidney Morgenbesser. More recently, I have learned from comments offered by audience members at my Rotman Lectures (University of Western Ontario), my Baker Lecture (Rice University), and my Teale Lecture (University of Connecticut). For suggestions about an earlier draft, I want to thank Gregor Betz, Marty Chalfie, Lorraine Daston, Anna Leuschner, and Torsten Wilholt. The detailed commentaries offered me by Allen Buchanan, Michael Fuerstein, and Evelyn Fox Keller have enabled me to make considerable improvements. Discussions with Michael over many years, and with Evelyn more recently, have been exceptionally valuable.

I dedicate this book to a wonderful friend and interlocutor: Michael Rothschild.

INTRODUCTION

Although it may not be a truth *universally* acknowledged, it is at least widely believed that the relations between scientific knowledge and the practices of the societies we think of as democracies are not what they should be. For some commentators, the problem is one of scientific pride, which seeks to override the values and the wisdom of the folk; for others, the trouble stems from public prejudice, which interferes with the proper translation of established knowledge into sensible policies. Whether the focal issues are the status of evolutionary theory, the proper use of molecular biology in medicine, the safety of genetically modified organisms, or the threat of global warming, people with starkly opposed ideas about what ought to be believed and what ought to be done concur in supposing the status quo to be unsatisfactory.

The chapters that follow treat these widespread attitudes as symptoms of a complex problem, that of integrating expertise with democratic values. The problem itself arises from many different causes: an oversimple view of inquiry, particularly scientific inquiry, as free of "values" and "value-judgments"; a conviction that fruitful discussions of values are impossible, and a consequent allergy to confronting questions about values directly; a superficial understanding of democracy that, in its crudest form, views democracy as achieved when people go to the polls and emerge waving ink-stained fingers in the air; a failure to understand that the system of public knowledge we have (in which the findings of the sciences occupy a specially privileged place), with its institutions and conventions, has emerged largely by happenstance, carrying with it residues of earlier states in which it was not expected to play anything like its present role. Only when these causes have been identified, and responses to them given, can one hope to offer a theory of the proper relation between expertise and democratic ideals.

I begin by reviewing some features of our current predicament, and by explaining why the picture of scientific research as a value-free zone is unwarranted. Chapter 2 then offers an approach for understanding value-judgments and for structuring debates about values. Summarizing an account I have offered at length elsewhere (Kitcher 2011a), I suppose people to have been engaged, for virtually all our history as a species, in the *ethical project*, a project in which we work out together how we should live with one another. Ethics is a social technology, one for which there are no experts. There is only the possibility of conversation, ideally free of factual mistakes and imbued with mutual sympathy.

Chapter 3 extends this general approach to values to offer a picture of democracy. Following Dewey, democracy is viewed as a work in progress, dedicated to expanding freedom through elaborating ways for satisfying mutual interaction. Besides the historically important problem of finding ways to resist identifiable oppression, typically visible in the person of a tyrant, people today must cope with the complexities of societies in which unidentified oppression can easily arise.[1] Ignorance, whether accidental or intentionally induced, can lead them to act in ways diametrically opposed to their central interests. Public knowledge ought to offer remedies for these predicaments.

In chapter 4, I trace the contingent evolution of systems of public knowledge. The current ways in which inquiry has been institutionalized grew out of the procedures of predemocratic societies and carry the traces of their past. Science as a community endeavor was not initially conceived as figuring in public life as it now does—it was not planned with its current part in mind.

On the basis of these four chapters, chapters 5–8 offer a theory of Science in a democratic society. Chapter 5 considers the context of investigation, arguing that the agenda of research ought to be shaped by the informed ideas of a broader public. Here I introduce the notion of well-ordered science (Kitcher 2001) and refine and defend it. Chapter 6 extends well-ordered science to the context of certification, the phase of inquiry in which new findings are accepted or rejected as part of public knowledge. It is important not only that certification be reliable but that it be *perceived* as being so. I diagnose various causes for the failure of transparency in certifying knowledge and suggest that these might be overcome through informed citizen involvement. Chapter 7 considers the difficulties that arise when an urgent problem seems to demand immediate action, before the community of experts is ready to speak with a single voice. I argue that debating these issues in a public forum, the "free discussion" for which people often clamor, is actually antithetical to the ideals of democracy, and that, once again, a better solution lies in incorporating the judgments of small groups of informed outsiders. Chapter 8 focuses on problems of diversity and dissent within the scientific community and among the broader public. It offers a rethinking of ideas about scientific credit, about the norms of research, and about the self-image of the scientist.

Chapter 9 returns to examples of the actual choices we face, considering how the framework developed in its predecessors might be used in settling debates about evolutionary theory, biomedical research, genetically modified foods, and climate change.

Although the discussion principally focuses on ideals for inquiry, there is also some attempt to offer suggestions for how we might move toward those ideals. Well-ordered science, especially in the extended form developed in chapters 5–8, is not something we are ever likely to obtain. Nevertheless, it is possible to envisage changes in our practices that would bring us closer to it. Typically, these involve advisory groups of citizens, representing a broad range of human perspectives, who are informed about the status of particular areas of research, and who serve as intermediaries between the research community and the broader public. As several commentators have seen, this is a natural way to give substance to the original idea of well-ordered science (Kitcher 2001), and it connects with the recommendations of those scholars who have supported a role for "citizen juries" or "deliberative polling" (Jefferson Project; Fishkin 2009). The approach offered in this book attempts to supplement these recommendations by offering a thorough analysis of the problematic relations between scientific expertise and democratic values, one that can define more precisely the various roles these juries might and should play. Whether or not my analysis succeeds, I hope it may identify some fundamental issues that should be addressed in any adequate account of how Science might fit into a democratic society.

Chapter 1

THE EROSION OF SCIENTIFIC AUTHORITY

1. DISAPPOINTMENT AND OVERCONFIDENCE

Since the eighteenth century, the forms of inquiry classified as natural sciences, sometimes collectively designated as "Science" (with a capital and in the singular),[1] have increasingly been viewed as the epitome of human knowledge, potentially capable of delivering great benefits for nations and for the human species. During the twentieth century, governments became convinced of the wisdom of investing in scientific research (Kevles 1978), and citizens became accustomed to think of natural scientists as authorities on whose conclusions they could rely. In recent decades, however, a variety of challenges to particular scientific judgments has fostered a far more ambivalent attitude to the authority of the natural sciences. Many Americans do not believe contemporary evolutionary theory offers a correct account of the history of life. Europeans are skeptical about scientific endorsements of the harmlessness of genetically modified organisms. Around the world, serious attention to problems of climate change is hampered by suspicions that the alleged "expert consensus" is premature and unreliable. The optimistic legacy of the Enlightenment is increasingly called into question.

It is easy to identify some sources of concern. Intellectuals who worry about the Enlightenment and its legacy may view particular academic movements as sources of the erosion of scientific authority. Reactions to fascism breed distrust of Enlightenment ideals and ideas (Horkheimer and Adorno 1978); philosophical ventures—under the loose label of "postmodernism"—disrupt the attribution of "knowledge" to scientists or "truth" to their judgments (Derrida 1976; Lyotard 1984); and historical and sociological studies of science insidiously depict scientific inquiry as politics by other means (Kuhn 1962; Foucault 1980; Bloor 1976; Collins 1985; Shapin and Schaffer 1985; Latour 1987). Reflection on these intellectual trends easily moves scholars who mourn the loss of trust in scientific authority to expose the errors of the supposed enemies (Gross and Levitt 1994; Sokal and Bricmont 1998; Koertge 1998). So the "Science Wars" are declared and fought with gusto.

Yet this has little to do with the real sources of concern. Skepticism about scientific authority has not grown because postmodernism has been injected into the drinking water. The movements briefly mentioned have various merits and flaws, but one common feature of them is a focus on *theoretical pictures* of scientific inquiry. Many of the authors who scrutinize historical, philosophical, and sociological judgments about the sciences are quite explicit in declaring that their critiques leave the status and authority of science in public life untouched (Kuhn, Shapin, Collins). Even the most widely read of them reach only a small fraction of the public, and most readers find the conclusions unthreatening: *philosophers* reacted defensively to Kuhn's ideas about the growth of scientific knowledge, but *scientists* who read his influential monograph have usually found his description of their practices both insightful and congenial.

Live skepticism about the authority of the sciences stems from a cluster of attitudes, far more prevalent in human societies. Particular *areas* of inquiry are suspect because of their perceived impact on ideas and ways of living that people cherish. Opposition to Darwin's ideas about the history of life persists because the evolutionary account is seen as denying central tenets of popular forms of religious belief—commitment to a literal reading of the Bible cannot allow that some major kinds of organisms were not created at the beginning. Citizens in countries with a checkered history of manipulating human lives, most notably Germany, are sensitive to the dangers of interventions using molecular genetics, and their concern fosters doubt about the reliability of the underlying biological ideas. The primary reservations are not about Science, in the singular, but about individual sciences, and, as the number of instances increases, areas that do not impinge directly on human lives and ideals are affected by contagion.

Yet these specific doubts are only one strand in the cluster. Alienation from scientific inquiry is reinforced by the sweeping declarations of enthusiastic scientists. For the legacy of the Enlightenment can be interpreted in two different ways. In its modest form, it claims only that certain types of scientific inquiry—first in physics, then in chemistry and geology, later in biology—offer reliable information about particular questions that matter to people; in the further course of investigation, there is hope for new findings in these areas and for the development of other fields in which similarly authoritative advice will be available. More ambitiously, the Enlightenment legacy may maintain that all questions of a specific type can be addressed by future research, or even that all important issues can eventually find scientific resolution. *Scientism*—to give the less restrained interpretation a familiar name—abounds, and its excesses stimulate reactions that detract from the credibility of more sober scientific judgments.

When the US Congress decided to fund the Human Genome Project, many supporters tacitly welcomed an opportunity to lead in an area of technology in which the United States would be competitive with the (then-dominant) Japanese, but the principal official reason for the initiative lay in confidence that mapping and sequencing the genome would yield cures for hereditary diseases. As many scientists and commentators pointed out at the time (Holtzman 1989; Nelkin and Tancredi 1994; Kitcher 1996), the immediate medical "benefits" would lie in increased powers to test and diagnose—providing knowledge that was not always welcome—and cures or methods of treatment would come only later, at a slow and uncertain pace. Nearly two decades on, genomics has enormously enriched our understanding of heredity, development, and cellular metabolism in a wide variety of organisms, as well as offering refined tools for evolutionary studies, but, as far as medical payoffs are concerned, the gloomy forecasts of the "pessimistic" commentators have proved, if anything, too hopeful. With a great deal of further research, most of it focused on nonhuman organisms, we may hope to understand enough about major human diseases to intervene successfully, but the beneficiaries of the program *begun* by the Human Genome Project are likely to be several generations removed from the politicians who enthused over the prospects of immediate success.

Scientism crept into the advertisements for medical genomics, overestimating the rapidity at which the promised benefits will accrue—yet the consequences desired may materialize one day. Far less certain are those claimed in the name of other widely touted areas of science. Those who celebrated the "decade of the brain" often believed that the development of neurophysiological tools and techniques, with a demonstrable record of success on a small range of relatively simple problems, could be expanded rapidly to fathom human psychology and behavior. Perhaps, centuries hence, historians of science will be able to look back to our times and identify the achievements of the pioneers who began a lengthy enterprise of understanding how the brain affects many important aspects of our lives. Or perhaps not. Even with the most dedicated efforts and the most ingenious investigations, the most fascinating facets of thought and emotion may prove too complex for us to comprehend.

For other forms of scientism, the gap between promise and performance is larger still, and yet more likely to endure. Since the 1970s, distinguished evolutionary theorists have argued that many topics traditionally investigated by the humanities and the social sciences can, and should, be tackled by deploying neo-Darwinian tools. Sociobiology ambitiously promised to sweep away the mushy and tentative efforts of psychologists and sociologists in favor of biological rigor. Contemporary evolutionary psychology has slightly tempered the original ambitions, but it remains reasonable to believe that the questions it sets for itself, if taken with the seriousness that pervades the most well-grounded evolutionary studies, are far beyond its resources. At its best, evolutionary theorizing provides precise models of the phenomena to be explained, and the development and testing of those models require details of genetics, development, and environmental conditions that are inaccessible for the forms of human behavior about which evolutionary psychologists spin their most celebrated yarns (Kitcher 1985; Vickers and Kitcher 2002; Haufe forthcoming).[2] Nor are matters better with another program of scientistic imperialism, one centered on the thought that the mathematical tools of neoclassical economics are sufficient to explain all human social phenomena.

Grand theories of nature-in-general, or of human nature, not only produce reactions of disappointment when they fail to live up to their advance billing, but they also contribute to a picture of science as an alienating institution. The thought of the community of scientists as embodying the modest reading of the Enlightenment legacy, patiently working, where they can, to provide valuable information that might assist this or that facet of human life, gives way to an image of a unitary movement, aiming to displace prior ways of thinking about the world and about ourselves in favor of a disconcerting novel understanding. For people who already worry about the judgments issuing from particular fields of research, ventures in scientistic imperialism easily provoke discomfort with Science, as a whole. Their resistance, or alienation, is reinforced by a third strand in the cluster.

The specific worries present particular sciences as a threat; scientism conjures the image of scientists as overambitious and arrogant. Prominent episodes in recent public discussions of various sciences suggest that scientific inquiry is inevitably prejudiced, biased by individual aspirations or political allegiances. Some scientists clamor for restrictions on familiar ways of behaving because of danger to the environment; others disagree, claiming any such constraints are unnecessary. Some researchers denounce products or procedures as unsafe; others contend the items in question are valuable and offer no serious risks. Newton's third law applies: to each scientist judging a controversy, there is an equal and opposite scientist. Even with respect to allegedly settled disputes—Darwin's account of the history of life, for example—there are "scientists" on both sides. Does it really matter that some belong to the National Academy of Sciences, and that others mostly teach at institutions set up to promulgate a particular faith? There are stories to be told to explain that distribution.

The "science-friendly" story is familiar. Part of the Enlightenment legacy consists in celebrating the freedom of scientific research from judgments of value: supposedly, there are value-neutral ways of evaluating evidence and coming to scientific conclusions. In cases of persistent controversy, some participants do not proceed in accordance with the value-free canons of evidence and good reasoning. Perhaps those who lapse are not skillful in these regards—they make mistakes in much the same way as those who bungle arithmetic. Or perhaps they are in the grip of antecedent prejudices, so blinkered they overlook important evidence or give a disproportionate weight to the fact that some problems remain, as yet, unsolved. The skilled, intelligent, and disinterested are rightly recruited to the most prestigious positions and are rewarded with the largest honors. By contrast, the unskilled and the biased make their careers at colleges, universities, and research centers where their specific conclusions resonate with the antecedent convictions of the founders and their contemporary heirs.

There is an alternative story. Institutionalized Science is dominated by people with biases that oppose the ideas of the folk. Behind the elite universities and the honorary societies is a subversive agenda, one intent on rooting out popular convictions and values. The teachers at the Bible colleges are debarred from other, more "prestigious," positions not because of lack of skill or "neutrality," but because their honest efforts run counter to the program favored by the elite. The story may continue by inverting the assessment of its "science-friendly" counterpart: the marginalized live up to the Enlightenment ideal of value-neutrality, and the "orthodox" are the ones infected by biases. Or, true to Newton's third law, it may concede that both sides make presuppositions, it may maintain that value-judgments are essential to research, and it may contend that there is no reason to sacrifice the rich values of the folk to the crass substitutes offered by arrogant scientific imperialists.

Recent decades furnish enough examples in which policies based on scientific judgment are publicly contested and in which citizens are not persuaded by claims of an "expert consensus" to substantiate my primary thesis that the authority of Science has been eroded. Attention to the rhetoric employed in these controversies reveals the cluster of themes I have viewed as responsible. Extensive sociological research would be required to justify thinking of these attitudes and concerns as the main constituents of resistance to or alienation from Science. Without being able to point to detailed data of this sort, I can only claim, tentatively, that a composite vision of the sciences as threatening, imperialistic, and resting on unsupported elite prejudices plays an important role. For my purposes, however, the incompleteness of the evidence does not much matter. Beginning with dissent about the authority of scientific "experts," and with possible, even likely, sources of discontent with Science, is a way of introducing the central problem with which this book is occupied.

2. THE DIVISION OF EPISTEMIC LABOR

One of the problems facing any democratic society is to decide how to integrate the plausible idea that, with respect to some issues, some people know more than others, with a commitment to democratic ideals and principles. An extreme way to solve the problem is to deny the plausible suggestion of unequal knowledge, or to deny that unequal knowledge matters. Policies for the society as a whole are to be thoroughly subject to discussion and vote: democracy requires, on this vision, that people make up their minds about the goals to be achieved and the facts pertinent to reaching those goals, free of any norm that would counsel trust in supposed authorities. Citizens are entitled to their own opinions across the board. Responsible participation in public affairs requires nothing further than making up your own mind, as you see fit. Call this *the commitment to epistemic equality*.

There is an alternative vision, or family of visions, of the ways in which democracies work. These perspectives favor the idea of a *division of epistemic labor.*[3] Consider the entire range of questions pertinent to public life, all the matters about what the society should aspire to and how it might realize whatever aims are set. These topics are partitioned, divided into nonoverlapping sets, and for each set in the partition except one, a particular group of people is designated as authoritative with respect to that set. For the remaining set, epistemic equality holds: that is, on these topics each citizen is entitled to make up his/her own mind. Because there are alternative ways to partition the topics, there are alternative perspectives that count as dividing the epistemic labor.

My abstract specification is readily supplemented with a comprehensible concrete example. You might think there are questions about what ideals to adopt, or what goals to pursue, and these should be assigned to that set of topics on which citizens should make up their own minds. Moreover some questions about facts might also belong in this set, issues about which each of us can be expected to be competent, or even privileged. Yet there are also areas in which people should defer to experts: many facts about distant regions might be assigned to the judgment of those who have been there, issues about the microstructure of matter seen as the province of particle physicists, hereditary phenomena recognized as belonging to the geneticists, matters of the functioning of particular machines awarded to the pertinent type of engineer, judgments about the authenticity of paintings to art historians, conclusions about the relations among various languages to historical linguists, and so on. On any vision like this, democratic ideals are honored because people are recognized as particularly able to identify their own aspirations and values, and the deference to experts is appropriate because those experts help them overcome the limitations of their knowledge, and thus to formulate and pursue their freely chosen projects more effectively.

To articulate the perspective more thoroughly, it would be necessary to introduce precise criteria to identify those belonging to the various expert classes. Exactly what range of experiences qualifies someone to pronounce on conditions in some distant region of the world? What must you have done to count as an expert on a specific group of diseases? Many of the disputes briefly discussed in the previous section arise out of dissatisfaction with a dominant division of epistemic labor, because dissenters suppose either that the criteria for attributing expertise have been misformulated or that adequate criteria do not apply to the people standardly designated as experts.

Some division of epistemic labor of the sort just envisaged is attractive precisely because of the difficulties of the polar alternatives between which it lies. On the face of it, the better information we have, the more likely we are to identify goals for ourselves and to fashion strategies for reaching them, both individually and collectively. Even under far simpler conditions than those reigning in contemporary democratic societies, each of us can benefit by learning from others; as we shall see in chapter 4, division of epistemic labor is almost certainly a very ancient human practice, profitably adopted as soon as we could speak to one another (if not before). Shared social life would be virtually impossible unless there were a common stock of information acquired by younger members of the community, and if the contents of this body of lore were entirely subject to the will of an adult majority, it would be easy to remain mired in conditions of dangerous ignorance. In the contemporary context, it is all too easy to appreciate how policy decisions on urgent matters can be stalled when all citizens think of themselves as equally capable of resolving factual questions.

It is useful to contrast the rejection of any form of expertise with the polar position that denies a division of epistemic labor by leaving everything to a small number of authorities. Plato's *Republic* offers a portrait of an allegedly ideal city, a *kallipolis*, in which the lives of all go well because wise and good experts understand what is best for each type of person and design institutions and laws that promote the best for all.[4] Plato seems to imagine that for each person there is a range of possible developmental environments (involving early socialization and education), and part of the task of government is to provide each citizen with that developmental environment leading, overall, to the highest level of social welfare, where it is understood that the (unspecified) aggregation function requires this to be done in a way that causes nobody to make large sacrifices of personal well-being. Thus the *kallipolis* contains citizens who are individually fortunate and whose lives harmonize to produce overall social good.

Plato knows the *kallipolis* will seem deficient to the friends of democracy, whose definition of what is good centers on freedom: "Surely you'd hear a democratic city say that this is the finest thing it has, so that as a result it is the only city worth living in for someone who is by nature free" (1992, 232). He is also well aware of the attractions of choosing a pattern of life for oneself: in a democracy each citizen "will arrange his own life in whatever manner pleases him" (1992, 227). Throughout his previous discussions, however, Plato has emphasized the bad consequences that flow when people try to do things for which they are unsuited, and he has tacitly supposed most of us have a tendency to do this (1992, 109). When he characterizes the ability to arrange one's own life as "licence," Plato is offering his verdict that, left to our own devices, each of us will bungle: our lives will go badly, and the life of our political community will suffer.

Many intelligent people find Plato's utopia repugnant. But what exactly is wrong with it? Do we assign a supreme value to pursuing the projects on which we have decided; is the possibility of self-direction so important to each of us that our objective good is bound up in it? Surely not. As things actually stand, we lack any extensive ability to chart the course of our lives: would it be preferable to sacrifice *all* the guidance and advice we receive in our youth? To the extent people in existing democracies make up their minds about what they want to do and be, their decisions are already constrained, and their minds are shaped by other people and by aspects of their environments beyond their control. If we are still vulnerable to blunder when left to our own devices, the loss of a little more self-direction comes to appear less consequential.

Those who detest the *kallipolis* do so not simply because they regard a power of self-direction as an overriding good, but because they harbor two important doubts about Plato's system. One is that the behavior of the rulers, Plato's guardians, is not likely to be directed toward the welfare of each and the welfare of all. The guardians will be corrupt, elevating their private concerns over the public good.[5] The other is that, even if they were impeccably focused on the public good, the guardians would not know the things Plato assumes to be available to them; specifically, they would not be able to identify the possibilities available to people (given various developmental environments), they would not be able to come to objective measures of individual welfare, and they would not be able to combine these measures into an evaluation of the general good.

The *Republic* attempts to block concerns of this type by proposing that there are some human beings who could attain the necessary knowledge if they were given a special kind of education (and the *Republic* goes into considerable detail concerning what should be part of this education and what should be excluded from it). One of the ingenious twists of Plato's metaphysics is to take the pinnacle of the educational process to be the recognition of the Form of the Good, so, as the guardians gain all the competences they need to rule, they also have to take on the incorruptibility that will silence anxieties about their abusing their powers: to be wise enough to govern is to become virtuous beyond corruption. None of this is likely to convince the champions of democracy, who will doubt that any selection of promising youth and any course of education we know how to devise could deliver the guarantees Plato assumes.

So, just as the commitment to epistemic inequality proves unattractive, there are sound reasons to reject the polar opposite of subjecting all decisions to those certified as Wise. Some form of division of epistemic labor seems called for. Yet my apparently unnecessary excursus through Plato's dystopia should give us pause. *How many of its repugnant features are present in the large-scale democracies we inhabit?* If the trouble with the *kallipolis* is that vast numbers of decisions are made for us (and we worry about whether the decision makers are wise and disinterested), the same is true of contemporary democracies. In our societies, the decisions are often made by short-sighted and venal people who have to defer to uncoordinated systems of expert advisers. Should we have any more faith that these decisions will be made for the general good? That those who make them are able to perceive what counts as the general good? That, to the extent they can perceive the general good, they are motivated to bring it about?

Of course, if contemporary voters object to what their representatives do, they can recall them. But we could easily amend the *kallipolis*. Imagine a system of education and testing, begun in early youth and continuing as long as you think appropriate, intended to mark out young people who are exceptionally good at solving complicated social problems, exceptionally wide-ranging in their capacity to acquire empirical knowledge, and noted for their probity and strength of character. (If you worry about how we could identify these characteristics, you should also wonder about our practices of choosing among students and of admitting people to positions of trust.) So, at the end of many years, a guardian class is formed, and at the beginning of each four-year period, a team of guardians is selected to pursue the administration of society. For four years, that team makes the decisions, consulting with others as the team thinks best. At the end of the period, the citizens can recall the team if they find the performance unsatisfactory. If the team is recalled, a completely new team is generated from the guardian class.

Do these modifications of Plato's program turn an unacceptable *kallipolis* into a satisfactory state? Surely not. Nor does the crucial difference lie in the fact that, in existing democracies, offices are open to all and votes are directed toward packages of proposals. In practice, the offices that wield effective power are open to a narrowly restricted class of citizens, and the packages offered to most voters leave them only limited, often unsatisfactory, choices. *As things stand, have extant democracies achieved a satisfactory division of epistemic labor, one that combines with social procedures to enable citizens to realize the goals democracy advertises as its own special virtues?*

The central questions pursued in the following chapters concern the character of a satisfactory division of epistemic labor, and how that division relates to deliberation and joint action in a genuine democracy. By considering two polar positions—the commitment to epistemic equality and the undivided authoritarianism of Plato's *kallipolis*—I have tried to isolate the question of the division of epistemic labor as an important issue. The challenge of differentiating existing arrangements from the fundamental character of the *kallipolis* indicates that we shall need to think about the promise, and the requirements, of democracy more carefully than I have yet done. Before turning to that task, however, I want to relate the points of this section to the questions of the erosion of scientific authority with which we began.

3. SOURCES OF TROUBLE

Public decision making about urgent issues and questions involving scientific complexities is often stalled because the "consensus of experts" is questioned: climate scientists who have spent decades trying to warn of the risks to our descendants are frustrated by the constant need to reiterate the explanations (Hansen 2009; Schneider 2009). To talk of expert consensus in cases like this is to invoke an *official* division of epistemic labor. Those who resist are moved by the thought that the official division is misguided or misapplied—it marginalizes dissenters with equal (or better) claims to expertise. Resisters may suppose that the marginalization results from prior prejudices, commitments to a particular set of values that should not intrude into scientific discussion. They view the result as a breach of democracy, a return to the authoritarianism celebrated in Plato's *kallipolis*, the silencing of open discussion that ought to occur.

This situation obtains because there is no satisfactory, well-articulated, and well-defended account of the proper division of epistemic labor and of its integration with the values central to democracy. *We urgently need a theory of the place of Science in a democratic society—or, if you like, of the ways in which a system of public knowledge should be shaped to promote democratic ideals*. The central task of this book—and, as I shall suggest below, the central question for the general philosophy of science (if not for epistemology) today—is to present a theory of this sort.[6] To undertake that task will require attending to general issues about values (chapter 2) and clarifying democratic ideals (chapter 3). Further study of the erosion of scientific authority will show just how these questions emerge.

Postpone, for the moment, the overridingly important debates about how our species should respond to the warnings of climate scientists, and focus on an example—far less significant except for its symbolic role—on which much ink has been spilled. Given the "official" division of epistemic labor, there is an *overwhelming* expert consensus on particular judgments about the history of life: our planet is over four billion years old; life emerged quite early in the earth's history; for most of that history, the only organisms were unicellular; all species are linked by relations of descent (in a large family tree—or bush); natural selection has played a large role in the modifications of organic forms. In several affluent countries, a significant number of people reject at least one of these claims. Many citizens of the United States repudiate *all* of them, and only a small minority would accept the "consensus view."

To appreciate the resilience of the resistance to the ideas typically associated with Darwin, you need only consider the limited impact of even the best efforts at biological education. During the 1980s, my former colleague at the University of Minnesota, Malcolm Kottler, taught the theory of evolution in a large lecture class, usually with an enrollment of about a thousand students. He devised some brilliant pedagogical strategies for explaining the central ideas—illustrating the persistence of useless traits by pointing to the odd residues of previous fashions in our clothes, having his students play a board game to show the workings of natural selection—strategies that I and others have shamelessly taken over in our own expositions. At the beginning of each semester, Malcolm gave each of his students a questionnaire, the data of which invariably showed that those accepting a Darwinian view of life were in a clear minority. When the same questionnaire was administered at the end of the course, he found some improvement, but nowhere near enough to turn the erstwhile minority into a majority: the increase was usually about 2 percent.

Because opposition to evolution ("evilyoushun") has been so carefully studied, its basis can be clearly recognized. Although the well-coached spokespeople who besiege their local school boards often cite particular challenges, supposedly revealing that "evolution is just a theory," the vast majority of those who want their children shielded from Darwinian doctrines are moved by more general considerations. According to their own view of the division of epistemic labor, there are some topics—crucially important topics about the cosmos, the human place in it, and the purposes of human existence—on which a particular group of people are well qualified to pronounce. Pastors, and the diligent students of the scriptures whose readings they lead, have insights inaccessible to the methods of secular science. The official division of epistemic labor excludes them from the community of experts on the history of life—and it even marginalizes those Christian paleontologists who seek to do justice to *all* the "sources of evidence," both whatever genuine evidence can be garnered in the usual scientific ways and the insights available only through the revealed word of God. To close the conversation by declaring a consensus, to shut out the voices of people who question the official views by patiently attending to the entire range of the evidence, is an affront to democracy. Some would even go further, regarding the insistence on stifling proper debate as proceeding from a "materialist-atheist" agenda that aims to subvert important values, and to deny the legitimacy of the beliefs by which many people steer their lives.

Two central themes emerge from this example. One concerns the particular way in which an official division of epistemic labor is questioned: whether because of the particular criteria adduced or through the application of those criteria, the class of "experts" in a particular field—the history of life—is narrowed, so that claims to knowledge are dismissed; the preferred explanation for the dismissal is that the claimants are victimized because of their general religious beliefs, subjected on that score to adverse, and inappropriate, value-judgments. The second theme focuses on the ideals of democracy: free competition among ideas is essential to democratic deliberation, and stifling debate is a violation of democratic values. As we shall now see, these themes are reinforced by the other examples of §1.

Consider the scientistic programs that advertise a future in which "serious" knowledge of human psychology, human behavior, and human society will be generated by deep understanding of our brains or our evolutionary history or our status as rational economic agents. Although we should celebrate the dedicated and ingenious work that has illuminated the neurobiology of sensation, sweeping claims on behalf of any of these programs typically portray the understandings of ourselves currently available to us as second-rate, if not confused and worthless. Resources that currently play a role in our deliberations, private and public, about our relations with one another, the contributions of various social sciences, as well as the humanities and the arts, are scorned, especially by writers convinced of the comprehensiveness of a scientific vision. Yet those are the resources we actually have for making important decisions. People already inclined to resist particular scientific conclusions, or to feel alienated from Science as a whole, will perceive the scientistic dismissal of areas of inquiry on which they draw as reflecting an arrogant distortion of the division of epistemic labor.

Manifestos on behalf of neuroscience, evolutionary psychology, and rational decision theory compete for the office of replacing the dim muddle of the traditional arts, humanities, and social sciences with the clear lines of scientific understanding. For all the successes with which a few aspects of human cognition and behavior—particularly facets of sensation—are now understood, the suggestion that we have a clear scientific vision of the most significant areas of human life is palpably absurd. Even if it is not the case that Sophocles and Shakespeare will *always* offer more insight into the human condition than all the scientists combined, there is no serious prospect that they will be outstripped, within the next century, by any army of researchers we might assemble.[7]

The ambitious thought that the natural sciences can explain and illuminate every aspect of ourselves and our world, that there are no modes of knowledge and understanding that are the province of artists and humanists, sometimes expressed in nonliteral language, sometimes even formulable only in different ways (by Beethoven or Vermeer, for example), is easily provoked by blind resistance to Science. Yet those who trumpet the comprehensiveness of Science often reinforce the blindness. In our times, sweeping scientism is often married to triumphalist atheism. Thoroughly convinced that the religions of the world offer a collection of noxious myths, eloquent champions of science propose to clear them away, inviting the hitherto benighted masses to join the great Darwinian party.[8] Loudly cheered by their supporters, scientistic evangelists polarize the discussion, intensifying the resistance to Science. They reinforce the image of arrogant dismissal of alternative claims to expertise, of a division of epistemic labor contrived to serve the interests of a prejudiced elite.

Back now to the debates about anthropogenic global warming. Once again, the discussions revolve around the proper division of epistemic labor. The public hears of an "expert consensus" on an expected rise in average global temperature of at least 2ºC during the current century, whatever we now do, brought about by human activities (specifically through the emission of greenhouse gases). It also hears—as, of course, in a democratic society, it should—that there are scientists who do not belong to this consensus, skeptics, naysayers, climate deniers. Why are these people excluded? Some, although eminent, have attained their honors by working in different fields, related only tangentially to climate science; others have ties to particular sectors of industry—they have testified for or worked in companies whose profits depend on allowing the present emission rates to continue. Should these attributes matter? What counts as a valid criterion of expertise, a proper division of epistemic labor?

Yet this debate is more complex, because a bare recognition of a rise in mean global temperature is hardly enough to guide in the framing of policy. What follows from the fact, assuming it is a fact, that temperatures will increase? Rising sea levels, almost certainly. But how many lives—and whose—will that endanger? What other risks will threaten people? To what extent will our descendants struggle to find shelter? Or suffer drought? Or face severe problems of famine caused by disrupted agriculture? Or encounter turmoil produced by the chaos of mass migrations? Or be vulnerable to new patterns of disease prevalence? Or be subject to plagues and pandemics, as novel disease vectors thrive in different environmental conditions? Given the official division of epistemic labor, there is no consensus on any of these questions. Some experts are prepared to offer estimates; others believe that even rough assessments of probabilities are unwarranted. Why, then, is there any need for preventive strategies now?

Opponents of immediate action have many options. They can question the base-level consensus, the division of epistemic labor that excludes the climate deniers. They can point to the lack of agreement about the foreseeable consequences. They can emphasize the political difficulties of concerted international intervention. They can propose that later generations will have a clearer view of what the genuine threats are, and that their increased wealth will enable them to respond more effectively than we can now. Combining all these responses, and adding a paean to democracy, they can call for further discussion, rather than some authoritarian decision that would put an end to the free exchange of ideas and override the legitimate opinions of many citizens.

Once again, the same themes: an insistence that the division of epistemic labor has been distorted, coupled to emphasis on the central role of open discussion in democratic societies. In this instance, however, an extra element, one less explicit in the other cases, emerges with particular clarity. Due deference to experts requires those experts to proceed according to objective standards: they should appraise the evidence dispassionately, uncorrupted by any judgments of value. Proper division of epistemic labor presupposes the value-freedom of Science. By this standard, the most prominent advocates of climatic intervention deserve indictment. In their writings—their "manifestos"—they overstep the bounds of their roles as scientists, importing their own, idiosyncratic, values. Because they have violated the conditions that frame their expertise, their urgings should have no authority.

In fact, the community of climate scientists is caught in a dilemma. If they record the range of opinion among them in sober prose, admitting their uncertainties, offering what probabilistic estimates they can, pointing out the range and complexity of possible consequences, their lengthy summaries cannot be expected to guide any swift action. Further evidence gathering, and further debate, will "surely be appropriate." On the other hand, if they are moved by the importance of responding to particular future scenarios, if they see the inundation of regions with millions of inhabitants as a serious risk and judge that consequence to be unacceptable, they will be accused of tainting their science with judgments of value. Under the conditions of value-free science, conditions enthusiastically championed by opponents of climatic interventions, they can only speak as scientists if they agree to be muffled.

Although it is the most significant, and most urgent, of the debates in which scientific authority is currently eroded, the controversy about climate change is hardly unique in raising issues about the entanglement of science and values. Effectively, the past decades have presented citizens with a sequence of examples in which dueling scientists could be viewed as importing values into the conclusions they drew from their inquiries. Those episodes have convinced many of those who have lived through them that scientific practice typically *does not* live up to the standard of disinterested inquiry on which its authority is supposed to rest. Why should anyone believe the warnings of tree-hugging, polar-bear-loving, antibusiness worrywarts—or, for that matter, the profit-seeking, free-market devotees who offer contrary advice? Why not simply follow one's own inclinations?

Any theory of the type with which I am concerned must come to terms with the sources of the erosion of authority. It must investigate what democracy requires, and it must consider how properly to assess, acknowledge, and limit expertise. A useful place to start is with the question that has just emerged. If scientific judgments are to have authority, must scientific practice be value-free?

4. VALUES AND SCIENCE

Insistence on the value-freedom of scientific investigation is common among scientists, journalists, and the general public. Once, it was also popular with students of science—historians, philosophers, and sociologists. More recent investigations of the practices of scientists, past and present, have revealed the dominant roles different types of values have often played in their judgments and inquiries. The scope of those investigations prompts the conclusion that this is more than unmasking, showing the Heroes to have feet of clay. As the instances mount, the ideal itself becomes suspect: it is not just that many individual scientists *do not* live up to the standard of value-freedom, but that they *cannot* do so—and therefore it is not the case that they *should* do so.

Start with a relatively obvious point about scientific practice. Investigators have to determine what problems deserve their time and effort. Any such decisions must turn on what contribution they might make: they must judge that some potential outcomes are more significant than others. That might be because of a general view that discovering the answer to a particular question—how to synthesize a particular molecule, say—might play a role in further investigations, ultimately helping to construct an overall picture of some range of phenomena, *where having a perspective on those phenomena is taken to be something worthwhile*. Some questions gain significance from the potential value of tools for further experimentation: if you can insert a fluorescent molecule into a particular family of cells, you have a way of tracking important developmental processes.[9] Others gain significance because answers would lead, relatively directly, to improvements in human well-being. The important point for the present is that *all* these assessments require judgments of value, for all identify a specific goal as worth achieving.

Of course, these general considerations are only the background against which researchers make their decisions about what to do next. Any sensible reflection on their own research has to take into account what kinds of problems are tractable, given the current state of the field—however valuable it might seem to gain an answer to a particular question, it would be quixotic to undertake projects for which the existing conceptual and material resources are plainly inadequate. Moreover, researchers have to ponder their own abilities and proclivities. Where, among the feasible ventures, can they best expend their own efforts? Further value-judgments can easily enter into the assessment of tractability and lead to the conclusion that a specific line of research is best suited to one's own talents. For present purposes, however, the important point is that general value-judgments about the significance of questions are inevitably part of the background against which the decisions are made.

Friends of the value-free ideal will surely greet these points with a yawn. Nobody ever thought, they will declare, that values could be expunged from all contexts of scientific decision making. Philosophers routinely distinguish between the context of discovery and the context of justification (Reichenbach 1949; Hempel 1966), recognizing the former as serendipitous and subject to all kinds of psychological-sociological-political-aesthetic considerations. The ideal of value-freedom is supposed to apply when scientists are deciding whether or not to accept some hypothesis (or theory) on the basis of evidence. Here, we want them to be guided by objective standards, not to acquiesce because they find the hypothesis appealing, or to reject it because they take it to threaten things they cherish. Values enter into scientific decisions about what questions to pursue—at the very beginning of the context of discovery—but that is utterly irrelevant to the ideal of value-free science, as it is properly conceived; to wit, as a constraint in the context of justification.

So neat a riposte would be adequate if considerations about future projects—lines of new research or direct applications—were never relevant in contexts of justification. Can scientists always make up their minds about the adequacy of the evidence without thinking about the ways in which acceptance or rejection of the hypothesis under scrutiny would affect the advance of science or the lives of people? Apparently not. Several decades ago, eminent philosophers of science debated whether scientists, in their assigned roles as scientists, can avoid value-judgments (Rudner 1953; Jeffrey 1956; Levi 1960). Those who denied that possibility pointed out very clearly that in many areas of scientific practice there are foreseeable consequences of being right or being wrong. If wrongly accepting a hypothesis would lead to outcomes judged to be disastrous, the standards for acceptance rightly go up. Conversely, if there are valuable results to be achieved if the hypothesis is correct, cautious insistence on "further evidence" can seem callous, even cruel—as it did to the AIDS activists who demanded an end to "blind trials" when they saw how well patients treated with the drug (rather than the placebo) were doing (Epstein 1996).

Advocates of value-freedom have a reply to this point. Distinguish the decision to accept a hypothesis from the attribution to that hypothesis of a particular probability. Strictly speaking, scientists in the context of justification are concerned with the latter: given the evidence they have, it is possible to provide an objective measure of the strength of evidential support, and, in principle, that is what they should do. On many occasions, the consequences of acceptance and rejection are remote or uncertain, so they are entitled to go further, to accept or reject, when the degree of evidential support (the probability of the hypothesis on the evidence) is very high. When there are foreseeable consequences, whose value, positive or negative, might be assessed, they should properly restrict themselves to reporting the degree of support, and leaving to the public (or representatives of the public) decisions about how to deploy the value-free information they have transmitted. In this way, the correct separation between value-free science and practical decision making is preserved.

This would be all very well but for a number of important difficulties. The first of these queries the idea of an "objective measure of evidential support." According to the most widely accepted formal account of scientific evidence, the probability assigned to a hypothesis depends essentially on the prior probability of that hypothesis (the chance that the hypothesis is true, given no information whatsoever)—*and the choice of prior probabilities is subject to no constraint except that they should be strictly between 0 and 1.*[10] Prior probabilities chosen at the whim of the investigator can be influenced by any type of value, with impunity, so the ideal of "value-free justification" lapses. Second, the very idea of attributing probabilities to hypotheses, whatever its attractiveness to some philosophers and methodologists, frequently seems ludicrous. That idea has its home in statistical contexts, where there are developed tools for assigning numerical values to the chances of particular outcomes (you can construct a plausible sample space). Elsewhere, it is simply baffling—as when one faces the question "What would be the probability of observed diffraction patterns if Fresnel's hypothesis of the wave character of light were not true?"

Consider the plight of a climate scientist, challenged to offer a precise probability—or even an interval within which the probability lies—for the consensus claim that anthropogenic global warming will issue in a mean global temperature increase of at least 2ºC during the next century. What would be a responsible answer? That the probability is "very high"? Or, even more poignantly, consider the predicament of the supposed Cassandras, who have deserted the proper standard of value-free science to declare that the evidence for climate change is sufficiently strong to require immediate action. How would we force them to substitute a *scientifically responsible* judgment for the conclusions, so obviously (and supposedly wrongly) permeated by values, they actually make? "Tell us," we might say, "the exact scenarios you envisage, and the exact probabilities you assign to each of them. That is the value-free information we need, and on that basis, *we* (the public) can consider the values to be assigned to the outcomes and thus resolve what to do." Our worried informant could only respond along the following lines: "I know there are lots of future possibilities I cannot foresee, but there are some potential outcomes I can. They are based on climatic models whose realism I cannot assess with any great precision. There's a whole spectrum of dire outcomes, involving direct loss of human life, vast disruptions of human communities, drought, widespread loss of shelter, breakdowns in agriculture and famine, new patterns of disease incidence, evolution of new disease vectors. Individually, these outcomes have chances I can't estimate. Collectively, I think there's a significant chance of a combination of extremely bad consequences—the chances being sufficiently high and the results sufficiently bad, that we should act now to reduce the likelihood events like these will occur. I'm sorry, but I can't do any better; I can't give you the atomistic breakdown of probabilities assigned to events you seem to think a responsible scientist should provide." Faced with this frank explanation, is it really appropriate to dismiss our scientist as an expert informant because of our commitment to the ideal of value-freedom? Do we really think *our* judgment—or that of anyone else—would be as good as that of a scientist whose lengthy immersion in these issues leads to the admittedly imprecise assessment offered?

Yet more serious than either of the difficulties so far raised for the attempt to maintain science as a value-free zone is the recognition of the ways in which value-judgments are deeply embedded in the practice of science. There is a strong tendency among commentators on science to oversimplify the work that precedes the endorsement of a hypothesis, to think of the scientist's efforts as neatly divided into a phase in which a line of research is chosen, a stage in which evidence is gathered, and a period in which the evidence is assessed, when the investigator reaches an objective judgment about the degree of support garnered by a hypothesis. In all but the simplest instances, however, simple sequences like this are multiply iterated. At various stages, the researcher has to decide whether what has been done so far is enough to warrant taking the next step. If the goal is to explore the role of a particular molecule in intracellular metabolism, for example, it will be necessary to decide if the substance you plan to inject is a sufficiently pure sample, if the cells into which you have inserted it have taken it up in the expected way, if unwanted side effects have been eliminated, and so on. At each step, the decision to go on involves a determination that the tests and probes you have made provide a basis for pursuing your next goal. Moreover, the entire course of the research is typically not foreseen in advance. Your goals adjust and evolve as you encounter unanticipated difficulties. New questions arise as worthy of investigation. Value-judgments are constantly made, and the investigation cannot be reduced to some neat division of contexts that allows values to be factored out at the end. (Douglas 2009; Wilholt 2009).

In this respect, most of everyday scientific practice embodies just those features of large historical changes that have challenged simple ideas about the rationality of Science (Collins 1985). Formal approaches to scientific confirmation are unable to provide any plausible reconstruction of major revolutionary episodes (Kuhn 1962; Feyerabend 1975; Kitcher 1993). In these transitions, the issue is not one of finding a measure of the evidential support generated by a *consistent* body of evidence, but one in which there are plausible, but incompatible, claims at all levels, partial problem solutions here and difficulties that resist solution elsewhere. The deep phenomenon behind what Kuhn and Feyerabend labeled "incommensurability" is the lack of any common measure on which the partial successes and apparent failures of rival complex bodies of broad doctrine can be assessed. In contexts like these, you can play with probabilities, assigning them to generate whatever conclusions you want, but because there are no constraints to separate the reasonable choices from the unreasonable ones, nothing useful is revealed.

That does not mean the history of science should be viewed as a sequence of irrational transitions, in which scientists lurch from one unsupported perspective to another. On the contrary, the decisions taken in the large revolutions, like those occurring daily in laboratories all over the world, are eminently reasonable. The point is that these cannot be reduced to the simple formalisms often taken as constitutive of rationality. *Reductive programs fail because they are insensitive to the—reasonable—value-judgments pervading scientific practice.*

There is a strong temptation to believe that to show a decision is aligned with canons of good reason and good judgment requires producing explicit canons and demonstrating how the decision exemplifies them. To resist the temptation, we should recall the many complex instances in which good judges are unable to articulate precise rules that guide them—detailed immersion in complex legal decisions can bring conviction that the judgment was a good one, even though there is no body of explicit theory to which one can appeal to support the conviction. We should think of the resolution of complex scientific decisions in a similar fashion. Perhaps "rational resolution" is a family resemblance concept, one we garner by clear perception of instances and foils; or, if it does allow an informative explication, perhaps the explication will come, as with so many concepts of scientific importance, only much later in inquiry.

One way to understand the reasonableness of scientific decision making is to immerse yourself in the details of the cases (Rudwick 1985; Kitcher 1993, chaps. 2, 7). For major historical transitions, it is not hard to discern a common structure. At early stages, the rival participants appeal to different successes and acknowledge different challenges. Their defenses of their perspectives rest on the adoption of different *schemes of values*: each side claims its accomplishments are the really crucial ones. Rival schemes of values may identify different values, or, probably more usually, they may focus on the same list of values but weigh them differently. When schemes of values clash, the subsequent course of scientific dispute consists in each side's trying to extend its own range of successful solutions, while making trouble for the other. As this occurs, retention of one of the doctrines can easily require modifying the scheme of values—you cannot continue to insist that *these* are the really crucial problems, as your opponent starts to find defensible answers to some of them. In this way, *commitments to factual claims and to value-judgments coevolve*. To appreciate the reasonable resolution of a dispute is to show how the process culminates in a situation in which there seems to be no coherent scheme of values for the losing side to adopt. For example, the resolution of the chemical revolution was achieved as Lavoisier and his phlogistonian opponents attempted to find a consistent set of representations of an ever-widening set of known reactions, and, as they proceeded, the scheme of values for phlogiston chemistry was forced in ever more peculiar directions (Kitcher 1993, 272–90).

The picture I have been outlining provokes an obvious concern. Resolution turns on showing how the losers are committed to an untenable scheme of values. Reflection on the diversity of values within the societies in which contemporary science is advanced can, and should, inspire concerns about whether a scheme of values can *ever* become untenable. Perhaps all, or virtually all, of the scientific community will eventually find it impossible to endorse a scheme of values that accommodates a particular approach, but matters may be quite different when the value-commitments of outsiders are considered. Might it not turn out that there are doctrines now widely, even virtually universally, accepted by scientists, whose acceptance would *not* have been licensed by a majority of the members of their societies, even when those members were fully and thoroughly informed about the details of the successes and failures that once led and now lead to scientific acceptance?

At this stage, the vague talk of "value-judgments" or of a "scheme of values" needs more precision. I propose a threefold division. A common conception, when Science is not the focus of discussion, takes a scheme of values to be a set of commitments around which someone's life is organized. People have ideals for themselves and for their societies, goals they take to be of the first importance and others they pursue but regard as subordinate. Call this a *broad* scheme of values, to recognize the wide scope it has across many dimensions of human lives.

Part of someone's broad scheme of values may be a concern for obtaining knowledge or for the attainment of knowledge by the society to which the person belongs. Particular kinds of knowledge may be valued for their own sake, or because they would be expected to form the basis for important applications and for the solution of problems the broad scheme of values marks out as important. So, to take an obvious example, if someone believes ending global poverty is an important ideal for human beings to strive toward, that person may view questions in molecular genetics as important because answers to those questions would allow for the development of drought-tolerant crops, for the provision of regular supplies of food to people who face the recurrent threat of starvation, and, ultimately, for the reduction of worldwide poverty. Call this kind of scheme of values a *cognitive* scheme of values; it represents the person's commitment to the ideal of gaining knowledge and marks out the kinds of knowledge the person takes to be especially important.

The third type of scheme of values is that most pertinent to the study of large transitions in the history of science. Within a major scientific controversy, rivals often share a cognitive scheme of values—the participants in the chemical revolution agree on the importance of knowing what is occurring in a certain set of reactions. Nevertheless, the disputants disagree with respect to their *probative* scheme of values. One party holds that addressing certain specific questions is crucial; their opponents have a different set of preferred problems they think should be resolved.

It might seem as if this third type of scheme would be entirely derivative from the cognitive scheme of values. For the probative scheme of values might be identified with that selection of problems that, given the person's beliefs, are taken to be the most reliable indicators of success across the entire range of issues marked out by the shared cognitive scheme of values. This, however, would be to ignore the ways in which the schemes of values can interact with one another. One does not have to conform one's probative scheme of values to the apparently most reliable way of satisfying the cognitive scheme of values.[11] Rather, the cognitive scheme might be revised under pressure from the probative scheme or under joint pressure from the probative and the broad scheme—just as the broad scheme might be revised under pressure from the cognitive scheme. The existence of possible tensions among these schemes, and of theoretically possible ways of responding to those tensions, bears on the questions about Science and democracy that are central to this book.

Suppose, for example, that your current cognitive scheme of values picks out a particular kind of knowledge as worth having: in the interests of achieving some goal (marked out by your broad scheme of values), you view it as important to find a model of some range of phenomena, precise in particular respects, accurate to some particular degree. Existing attempts to find a model of the type envisaged, when assessed by your probative scheme of values, prove defective. As you continue to explore the possibilities, none of them satisfies your demands. At this point, you might relax the standards of reliability, adjusting the probative scheme of values so that it is more liberal. Or you might conclude that the most general ends you wish to achieve, endorsed by your broad scheme of values, must be pursued in some different way—you modify your cognitive scheme of values to single out some different type of knowledge that would enable you to reach your goals by a different route. Finally, you might come to think the broader goals you have set for yourself are unattainable, amending your broad scheme of values, as, for example, when scientists gave up on the thought that the pursuit of inquiry into nature would reveal the wise purposes of the Creator.

Let us take stock. Because of the intricacy of most scientific research, value-judgments are deeply embedded in scientific practice. Scientists rely on their *probative* schemes of values to decide when they can continue to the next stage: they judge that the setup passes the *appropriate* tests for them to conclude that things are as they intended. They adjust their goals in light of their *cognitive* schemes of values, and, in some instances, in light of their *broad* schemes of values: as the data come in, you can appreciate the possibility of achieving a direct social benefit you had not initially envisaged. Moreover, individual scientists take for granted large perspectives that have often emerged from historical transitions in which rival schemes of values were pitted against one another, episodes in which factual beliefs and value-commitments have coevolved. If these assessments are correct, there is no possibility of defending Science as a value-free zone.

Nevertheless, it is easy to appreciate why the value-neutrality of scientific inquiry appeared so important. For there are incursions of values many champions of Science would regard as highly unreasonable. The debate over the history of life provides a compelling example. Darwin's opponents are often committed to a broad scheme of values that emphasizes the importance of belief in the inerrancy of the scriptures. Derivatively, their cognitive scheme of values attaches overriding importance to showing how geological and biological findings are in harmony with their reading of *Genesis*. In consequence, they will adopt a probative scheme of values denying the importance of what orthodox evolutionary theorists take to be the massive evidence in favor of their principal claims. On what grounds can their stance be diagnosed as flawed?

Those sympathetic to the value-free ideal might reasonably claim that, even though it may be unattainable in practice, inquiry goes better the more closely it can be approximated. So scientists might try to minimize the value-judgments they make, or try to derive the same conclusions using different clusters of value-judgments.[12] I agree with the thought of viewing freedom from value-judgments as a standard we might do well to approximate, when and to the extent we can. Yet I want to resist the suspicion that the incursion of values inevitably undermines scientific authority. Reliance on *some* values is genuinely troublesome—witness the example of opposition to Darwin, noted in the previous paragraph—but there are value-judgments and value-judgments, and some of them are unproblematic.

If value-judgments pervade scientific practice, how can the authority of scientific judgments be maintained? The question presupposes a popular idea, one that leads scientists and commentators on Science to uphold the value-free ideal, and to shy away from admitting to judgments of value. At bottom, the concern is that value-judgments are subject to no standards at all, that valuing is a matter of taste, that where value-judgments come in, intellectual chaos is not far behind. An allergy to public value-judgment has long pervaded our discussions of Science, fostering the myth of some neutrality that is actually attained. The deepest source of the current erosion of scientific authority consists in insisting on the value-freedom of Genuine Science, while attributing value-judgments to the scientists whose conclusions you want to deny.

A very large part of the counsel offered by scientists should be authoritative in our public deliberations, even though value-judgments figure in their decisions. It is entirely possible that the advice of my imaginary frank climate scientist deserves to be followed. Everything depends on the character of the value-judgments made, on the schemes of values and how they are applied, on the ways in which schemes of values are adjusted. If there are standards for endorsing, applying, and amending values, we can separate the religious opponents of evolution and the hired guns of industry from the genuine experts, appropriately distinguished in our division of epistemic labor.

A theory of the place of Science in a democratic society needs to start with a perspective on values and valuing. Instead of "Science and Values"—as if the latter were grudgingly tacked on as an afterthought—we should think in terms of "Values and Science." The next chapter attempts to introduce the required perspective.

Chapter 2

DISCUSSING VALUES

5. ETHICS AS A HUMAN PROJECT

Despite the allergy to recognizing the role value-judgments play in the growth of human knowledge, most people do not endorse the idea that any judgment about what is good or bad, right or wrong, valuable or not, is as defensible as any other. The overwhelming majority of members of most human societies typically suppose some judgments are privileged, and that the source of the privilege is the will of a supernatural being, who has revealed important decrees to their ancestors. Ever since Plato, philosophers have been skeptical about the proposal that the practice of valuing could be grounded in the divine will. Many great thinkers have offered rival ideas about the objectivity of values, although none of them has been sufficiently convincing to replace the popular conception of a theological foundation. I shall sketch a different account, one that invokes neither deities nor other sorts of nebulous entities and processes: the moral law within, faculties of practical reason, perception or intuition of the good, an abstract realm of values, a distinguished class of moral sentiments. There will be no spooks.[1]

Valuing emerges from a human practice with an extremely long past, from our involvement in the *ethical project*.[2] That project began tens of thousands of years ago, in response to the difficulties of our lives together. Based on the findings of contemporary anthropologists, I suppose our ancestors lived in relatively small bands—perhaps fifty members—mixed by age and sex; their social situation was akin to that of our closest surviving evolutionary relatives: chimpanzees and bonobos. To live together in this way, they required a capacity for psychological altruism. That is, they had to be able, on occasion, to respond positively to the perceived wishes and intentions of those around them, to recognize what another band member was attempting to do, and to modify their own plans so as to help the other realize a goal. Despite the perennial popularity of conceiving animals (including human beings) as inevitably egoistic, there is strong evidence for attributing an altruistic capacity of this kind (Goodall 1988; De Waal 1996; Kitcher 2011a, chap. 2).

Psychological altruism made a particular type of social existence possible for our ancestors, but it would be wrong to suppose that living together in this fashion was easy. Observations of contemporary chimpanzees and bonobos reveal that the capacities for responding positively to others are very limited. Even animals disposed to attend to one another's manifest wishes, and to act cooperatively as a result, sometimes fail to respond or even thwart one another's desires. Psychological altruism is multidimensional, varying in intensity for different partners and for different situations. When egoistic rewards appear very attractive, even an apparently reliable ally will desert a friend. The daily life of chimpanzee troops reveals many occasions on which altruism fails, often requiring lengthy sessions of mutual reassurance before the rents in the social fabric are rewoven. Hominid and early human societies, like the chimpanzee societies of today, were almost certainly tense and uneasy, marked by time-consuming activities of peace making and only limited cooperation. If that had been the fate of our own species, vast numbers of features we take for granted in our lives would have been absent.

Something changed, permitting us to cooperate more extensively, to engage with ease with people with whom we have never previously interacted, to live in far larger groups, and to build up a vast array of roles and institutions that presuppose complex coordination. What could it have been? Clues are supplied by the most ancient written documents that have come down to us, among which are collections of rules. The tablets are not, strictly speaking, codes of law, for they obviously presuppose a body of social prescriptions to which the rules they contain are being appended. Those rules are additions to a previous practice, one that has enabled thousands of people to live in the same place and to orchestrate their activities, and the new precepts are the latest responses to the most recent problems. Behind them stand practices of rule giving and rule following, extending back into the Paleolithic, practices that have successively made possible a distinctively human form of social life, one that allows for cave art, for trade, for larger societies, for burial rites, for the domestication of plants and animals, for agriculture, and ultimately for hierarchical cities with massive buildings and sophisticated systems of irrigation.

How could our species have come from there—the limited social condition of chimpanzee-like hominids—to here—the complex life of Mesopotamian and Egyptian cities? Through acquiring a capacity for *normative guidance*. Remedies for the altruism failures that constantly disturb the peace of animals, altruistic enough to live together but not sufficiently altruistic to avoid pursuing selfish temptations, were provided by the ability to recognize regularities in behavior that cause trouble, to formulate a directive for conduct, and to follow that directive. The ethical project began when our forebears learned to restrain themselves from courses of action that would generate unwelcome retaliation. Restraint was recommended by prudence and fear: aware of the possibility of "punishment," in the rough-and-ready sense in which punishment is inflicted in chimpanzee societies, the first ethicists checked the tendencies that would previously have led to altruism failures. Later in the evolution of the ethical project, different sorts of motivation would enter: emotional attachments to others and to the group; respect for the rules prescribed and for the practice of rule giving; a sense of awe and reverence as those rules became associated with the will of an unseen being, capable of supervising human conduct even when we think of ourselves as most alone.[3] Over tens of thousands of years, human beings have acquired a miscellany of psychological capacities underlying our self-regulation.

From an early stage, normative guidance was *socially embedded*. Like contemporary hunter-gatherers, our Paleolithic predecessors sat down together to decide on the precepts for governing their group life. Their discussions almost certainly took place on the egalitarian terms recognized in those small societies today whose conditions of life are most akin to theirs (Lee 1979; Boehm 1999; Knauft 1991): all adult members of the band are to be heard, and the wishes of each must be considered. To diverge from egalitarianism of this sort would risk the survival of the group, for all had to pull together on occasion, to meet the challenges of the environment. Ethics emerged from human conversation, evolving as different groups tried out alternative prescriptions and as bits and pieces of rival codes were transmitted from generation to generation, from band to band. A sprawling genealogical tree has its roots in the ventures of the pioneers, and the tips of its branches in the codes that survive in contemporary societies. Our practices result by a—long—series of modifications from the first, primitive responses to the tensions of chimp-hominid social life.

So far, I believe, the account sketched can be justified by the evidence available to us (evidence from psychology, primatology, archaeology, anthropology, and evolutionary theory). If, however, my promise to have avoided all "spooks" is to be honored, the early ventures in normative guidance—most probably rules for sharing scarce resources and against initiating violence, effective because of fear of crude forms of punishment—must have given rise, by a sequence of comprehensible steps, to the far more sophisticated forms of ethical life visible in the ancient world. How did notions of particular virtues emerge? How did a concern for avoiding social trouble produce ideals for one's own conduct and development, independent of the effects on other people? We can be relatively confident about the origins of the ethical project, and, with the advent of writing, we can understand its contours about five thousand years ago. Yet the clues from the Paleolithic and early Neolithic are so sparse that it is impossible to justify an account of how ethical practice was *actually* transformed.

But we do not need to know the *actual* course of events. Worries about my narrative arise from suspicions that nothing whose origins lay in crude regulation of social life *could* metamorphose into the sophisticated conceptions already present in ancient Egypt and Mesopotamia, and even more evident in Plato and Aristotle, unless something mysterious—nebulous, "spooky"—happened in the interim. Gradual evolution is impossible: some moment of revelation, intuition, awareness of practical reason or the moral law within is required. That is the challenge—and it can be turned back by outlining a *possible* sequence of events that could have led from the primitive beginnings to the sophisticated terminus. The outline can be given. It does not pretend to record the *actual* history, but rather to simply rebut the skeptic who alleges that no such history is *possible*.

Here is a bare sketch.[4] Rules for sharing were almost certainly introduced early in the ethical project—for failures to share are prominent breakdowns of altruism that infect chimpanzee social life. Those rules lead naturally to accepting the goal of securing enough resources for all. The drive to increase the group's stock of resources can motivate a division of labor, in which activities are partitioned according to perceived aptitude. Directives to obtain a surplus can be motivated by recognition that times are sometimes hard. Further awareness of the different successes of neighboring bands can inspire expansion of the division of labor, the beginnings of peaceful interactions and trade with outsiders, and a corresponding extension of some of the local rules to cover the trading partners (at least in limited contexts). Expansion of cooperative ventures, both within and among bands, prompts a more fine-grained assignment of individuals to particular activities, generating a conception of roles that demand particular qualities and that are subject to specific precepts. Ensuring that the more complex and demanding roles can be filled introduces rules for individual development and an emphasis on qualities the young should acquire. Animal domestication provides resources for the group, and those who are assigned to the most exacting roles may come to have greater influence in the disposition of the resources. Gradually, the initial equality gives way to a hierarchical arrangement, and the idea of group ownership is replaced by an institution of individual property. Throughout the entire process, repeated cooperation and practiced attunement of individuals to one another generates ideas of responsiveness and more complex forms of altruism. Eventually, the descendants of the original group can identify people as courageous and persevering, and they can formulate ideals of friendship and even love.

Once again, it should not be supposed that just this sequence of steps occurred in the transformation of the first pioneering efforts into the sophisticated ethical life of the ancient world. I claim only to have outlined, very briefly, how an ensemble of local communities, responding to difficulties and opportunities we can reasonably attribute to them, might have generated, gradually and without any nebulous moments of revelation or insight, the complex framework of ethical life with which we today are familiar.

6. ETHICAL PROGRESS?

The task of this chapter is to provide an account of the standards value-judgments should satisfy, and the previous section can easily seem to doom any such venture. If the ethical project originates and evolves through a series of conversations in which particular groups of people mull over the problems they encounter, how can its direction be anything more than happenstance, its long history any more than one damned thing after another? I have been at some pains to avoid invoking moments at which the sources of standards manifest themselves. There are no lonely encounters on mountaintops in which an impressive being turns over some large slabs of stone and dictates the words to be engraved upon them, no episodes in which a carefully educated philosopher enjoys a vision of the Good. How could the ethical project, as just conceived, furnish the standards we seek?

It is useful to start with a more precise question. Do ethical practices ever make anything that could reasonably be called progress? Even in the episodes already discussed, there is pressure to see the possibility of progress. The hypothetical extension of intragroup rules to apply to the neighbors, begun in the practice of trade, looks like a progressive step, for it seems to be an advance when the previous mayhem of interactions with them gives way to more orderly treatment. If we trace the historical development of ethical ideas, through periods for which we have written records, there are further instances, cases it is hard to resist seeing as progressive transitions.[5]

Among the earliest of the legal codes that have come down to us are prescriptions about what should be done in response to particular crimes. From the Mesopotamian perspective of four thousand years ago, a proper response to the murder of someone's daughter (son) is to kill the daughter (son) of the murderer; if a woman is raped, the victim's father should designate someone to rape the wife of the perpetrator. These recommendations embody a bizarrely literal version of the *lex talionis*: wrongdoers are to be punished by inflicting on them the same loss they have caused to others, even when the "vehicle" through which the punishment is administered is another person. A thousand years later, the apt intervention is conceived differently—now the murderers and rapists are directly affected; the murderer's life must be given for the life of the victim. This familiar conception may not be the last word in proper reactions to horrible deeds, but it initially appears an advance on the view it replaces. How could it not be ethical progress to give up harming innocent bystanders in favor of inflicting pain on the agent responsible?

A second example derives from a better-known chapter in Western history. New world colonies reintroduced a practice, common in the ancient world, but one that had lapsed in the countries from which the colonists had come. Chattel slavery, the ownership of people, was revived, and the unfortunate victims were subjected to the middle passage, the lash, branding, rape, and the separation of families. During the eighteenth century, opposition to slavery in the British colonies of North America mounted, and eventually the practice was abolished and condemned as wrong. Once again, it is hard to view this as anything other than an advance. Is it convincing to suppose the two transitions—first the reintroduction of slavery and then its abolition—really to be on a par?

Pressure to talk of ethical progress arises in cases like these and prompts a search for the standard, the external source, that those who proposed the change and campaigned for it must have recognized. Yet there are no easily identifiable moments of ethical discovery, analogs of those episodes in science where an investigator apprehends something new and unexpected (Röntgen's fluorescing screen, the background radiation of the cosmos, retroviruses). Sometimes the causes of ethical change are elusive because we know so little about when the transition occurred and the people who figured in it: we can locate a *millennium* in which the *lex talionis* was modified, but that is hardly to pinpoint the event! Even when our sources are far better, however, as with respect to the abolition of chattel slavery (or for the increasing acceptance of same-sex preference during recent decades), no moment of discovery can be found (Kitcher 2011a, chap. 6).

To understand the concept of ethical progress, we need to break free of presuppositions that confine our vision. Instead of supposing ethical progress consists in the accumulation of ethical truth—where the sources of truth are independent of us and our discussions with one another—we can take progress as the prior notion. There are areas of human endeavor in which we conceive progress independently of attaining truth: technological progress consists in introducing and refining devices that help us to overcome practical problems. Ethics can be approached in similar fashion, viewed as a *social technology* that liberates us from the difficulties of a human predicament. Our limited capacities for psychological altruism make a particular type of social existence possible for us, but the limitations of those capacities interfere with the smooth pursuit of that form of life. The *original function* of ethics was to solve the problems caused by the incompleteness and unreliability of our altruistic tendencies, to remedy some of the failures of altruism that restricted the social possibilities for our hominid forebears and that continue to confine the lives of our evolutionary cousins.

Technologies respond to problems, but partial solutions typically generate further problems. Mechanized vehicles enable people to travel farther and faster than they were previously able to, but once the new devices are widely employed, a whole range of novel issues must be addressed. Use must be coordinated and regulated; there have to be highways and airports, traffic lights and control towers, systems of training, insurance, and policing, and a host of further innovations. So, too, in the case of ethics. Responses to the initial problem of remedying altruism failures—a deep problem, central to the conditions of human life—generate further problems of expanding the resources for the group, of interacting with outsiders, of encouraging and rewarding those who are to play important roles. The attempts to address these problems produce people whose sense of a worthwhile life is far richer than any conception available at the beginning, people who can pose the question "How to live?" in the terms favored by the philosophers of the ancient world. Just as there is no ultimate transportation system, toward which our technology is tending, so progress cannot be measured in terms of our proximity to it; there is no ideal ethical system we are struggling to reach. Progress, in both instances, is progress *from*, measured in terms of our ability to solve problems—both the original problem that set the enterprise in motion and those that emerge from our responses.

When ethical progress is conceived in this way, it is possible, after all, to talk of truth in ethics. Truth does not, however, come first, discovered by those who make ethical progress. Rather, truth is constituted in terms of the tools that solve our problems. More precisely, true ethical statements are those corresponding to precepts that are introduced in progressive transitions and that continue to be maintained under an indefinite further sequence of progressive transitions: a rule enjoining honesty was almost certainly a progressive step in our ethical past, and something like it (albeit vague) will almost certainly be preserved in our future progressive steps. The truth of the statement "Honesty is good" *consists in* just that relation between the rule and ethical progress; truth "happens to an idea" (James 1984, 823). There is a core of imprecise statements on which all ethically progressive traditions can be expected to converge—but it is also quite possible that indefinitely progressing versions of the ethical project might never reach agreement in other respects.[6] The approach outlined supports the possibility of pluralism (Kitcher 2011a, chap. 7).

Pluralism arises from the possibility of functional conflict. As a form of technology generates new problems, compromises sometimes have to be made because solutions to one problem interfere with solutions to others—in the case of transportation, considerations of speed and safety pull in different directions. Often there are alternative ways of compromising, and rival traditions, fully aware of the desirability of overcoming a number of problems, set their priorities differently. They may progress indefinitely, continuing to improve with respect to all the desiderata yet never completely converge.

In the evolution of the ethical project, the egalitarianism of the early stages (almost certainly prevalent through most of our ethical past) has now given way, in almost all contemporary societies, to the acceptance of rank and hierarchy. Some small groups, whose ways of living remain close to those of the ethical pioneers, retain the emphasis on equal participation and equal treatment.[7] The obvious compensation obtained in the larger, hierarchical societies is the proliferation of human possibilities, a richer conception of worthwhile human lives. Here we find functional conflict, and the sense of difficulty in reaching ethical agreement—and the allergy to value-judgments it produces—stems from the suspicion that at this point nothing can be said. Rival traditions must go their separate ways, driven by different priorities.

The ethical project liberated our species from the difficulties of the chimpanzee-hominid predicament, and there is no alternative to continuing it. We are bound by the notion of progress characterized here, committed to the attempt to overcome the problems that have arisen, generated ultimately by the problem of remedying altruism-failures. Yet that commitment appears to allow for alternative ways of going on, of assigning quite different priorities to different problems, and thus generating radically alternative schemes of values. Does pluralism run rampant, creating just that intellectual chaos often taken to engulf discussions of value-judgments?

7. RENEWING THE PROJECT

In the beginning, there were no ethical authorities. Our ancestors who invented ethics did not make it up arbitrarily—they reacted to deep features of the human predicament—but none could claim an expertise not shared by others. Their social technology was forged in discussion, in which all adult voices were heard, and their goal was to find something with which all could be satisfied. Most contemporary people do not live that way. We think of ethics as a subject on which there are experts—typically spokesmen for a deity, although a few secular philosophers believe the authoritative role belongs to people who can apply some faculty of reason, or feeling, or perception, or intuition, or whatever, to deliver the truth. My perspective on the ethical project rejects that conception. The idea of ethical expertise was a distortion of the project, the outgrowth of an apparently potent device for increasing compliance to the agreed-on precepts: a supernatural enforcer is a wonderful idea for keeping people in line, but the transcendent policeman can easily be exploited by those claiming special access to his will. The distortion arose because it replaced the notion of an *authoritative conversation*, in which *all* participated on *equal* terms, with the concept of an external constraint to which ethical practice was to conform.

I *propose*—and the word is important, given the points just made—that we undo the distortion. The only vehicle available to us—to anyone—for arriving at judgments about values is discussion in which the participants come as equals, and in which the goal is to satisfy all (that is, to reach an outcome in which everyone can acquiesce). That leaves many things open. Who exactly is to be included? What constraints, if any, should be imposed on contributions? Here again it is only possible to make proposals, for there are no sources outside the discussion itself to which a philosopher—or any other would-be authority—can point. Proposals can be guided by an understanding of the past of the ethical project, they can be supported by showing how they address the problems the project has evolved to tackle, and they can be scrutinized for their coherence. There is, I maintain, a coherent package of proposals containing the following elements: (1) It takes the body of discussants to be the members of our species, including those who will come after us; (2) It supposes the discussion should meet conditions of *mutual engagement*; (3) It strives toward a situation in which all people have equal opportunities—*serious* equal opportunities—for living a worthwhile life. As we shall see in the next chapter, these proposals embody a rich conception of democracy.

The package of proposals can be viewed as an attempt to scale up, in the contemporary situation, the conditions prevalent at the beginnings of the ethical project and, indeed, prevalent through most of its history. Our deliberating ancestors overcame the tensions of hominid social life by remedying altruism failures, concentrating on those occasions on which members of the group were thwarted in obtaining those things that "made their lives go well"—in the minimal sense of making it "go" at all. Their bare notion of a worthwhile life was one dominated by the provision of very basic things. Living in a far richer world, one capable in principle of meeting the material needs of the entire human population, we can aspire to do more, to extend the egalitarian ideal of satisfying the most elementary wants to embrace the far richer set of desires made possible by the proliferation of human possibilities.[8] It replaces the small band by the entire species, recognizing the webs of causal interaction that link us to people who live at great distances from ourselves: if community is a matter of causal interrelationships, for many purposes we are now one community. As the first ventures in ethics responded to all members of the band—and treated them as equals—so these proposals reflect an egalitarian attitude to a vastly more extensive community.

Plainly, any close replication of the deliberations of the ethical pioneers is impossible: any attempt to orchestrate even a sample of voices representative of the diverse perspectives of living people would produce a vast cacophony, in which the divisions and distortions produced in our history would doom any chance of serious discussion. Instead, the ethical conclusions to be endorsed are those that would emerge from an ideal conversation, one satisfying the conditions of *mutual engagement*. These conditions form the core of my proposal for method in the investigation of values.

More precisely, our ethical discussions are adequate to the extent they reach the conclusions that would have resulted from an ideal deliberation under conditions of mutual engagement, and disclose those features of the ideal deliberation that would move participants to adopt that conclusion. Some of the conditions of mutual engagement are epistemic: participants must not rely on false beliefs about the natural world, they recognize the consequences for one another of the actions and institutional arrangements under discussion, and they are able to identify the wishes other participants have and how those wishes evolve in the course of the conversation. These epistemic conditions are important, because even the most well-disposed conversationalists will arrive at peculiar recommendations if they are in the grip of flawed ideas: a group of intensely altruistic people would arrive at bizarre conclusions if they assumed extremely severe pain has all sorts of wonderful consequences for sufferers.[9] Yet, as we shall see, the epistemic conditions rule out many of the firm pronouncements that actually disrupt conversations about values, including the familiar assertions that particular things are required because they are commanded by one's preferred deity (under one's preferred interpretation).[10]

The heart of my account of mutual engagement consists in *affective* conditions. Start from the thought that genuine engagement with others begins from an expansion of one's sympathies, in which the perceived desires of those with whom one deliberates are given equal weight with one's own. Because of conflicts, that cannot be carried out consistently across the board. If two people have incompatible desires, it is impossible for a third party to incorporate both of their preferences into her own altruistic attitudes. Nor will it do to seek identification with the wishes of some harmonious majority, for the majority may be blinded by failures of sympathy. How, then, is expansion of sympathy to be conceived?

I tackle this question by introducing the notion of *mirroring* others.[11] The simplest sort of mirroring is that just considered (and found problematic as a general account of mutual engagement). For *A* to engage in *primitive mirroring* of *B* is for *A*'s (sympathetic) desire to give equal weight to the (egoistic) desire of *B* and to *A*'s own (egoistic) desire. Now, in an ideal conversation aimed at addressing functional conflict in an ethical code, the solitary desires from which mirroring begins are not those individuals actually adopt, but rather are in harmony with the functions to which the ethical code responds and are sustainable if the cognitive conditions on ideal conversationalists were satisfied. Hence the egoistic desires of ideal participants must first be filtered to include only such desires as are compatible with precepts of the participant's ethical code (precepts contributing to the functions the code is to discharge) and as are also retained when the participant accords with the epistemic conditions.

Primitive mirroring cannot provide a general account of the affective part of mutual engagement, because *A* can encounter situations in which two others, *B* and *B**, have incompatible desires, so *A* cannot accommodate both of them. Given the required filtering, however, some of these problems can disappear, and primitive mirroring of others may become possible. When that occurs, an ideal participant adopts the pertinent desire (one that gives equal weight to the desires resulting from the filtering).

When differences remain, engagement with others proceeds through *extended mirroring*. In extended mirroring, *A* attends not only to *B*'s (filtered) desires but also to *B**'s (filtered) assessment of *B*'s (filtered) desires, *B**'s assessment of *A*'s assessment of *B*'s desires, *B***'s assessment of *B*'s desires, *B***'s assessment of *B**'s assessment of *B*'s desires, and so forth. Through consideration of a variety of perspectives, a conversationalist seeks the best balance among the ethically permissible and factually well-grounded desires present in the population.[12] Ideal conversationalists form their sympathetic desires by extended mirroring of the desires of others, achieving the desires they judge to be the best balance among the varying assessments (indefinitely iterated) made by fellow participants.

If there is complete agreement about how the balancing is to be done, there is no need for further conversation. If there is not, the ideal conversation consists in attempts to support or reject various ways of balancing. Those attempts seek to show that proposals participants desire to implement as ways of responding to functional conflict either accord or fail to accord either with ethical functions all participants recognize or with their shared understanding of the need to respond to the wishes of all.

Ethical discussions addressing disagreements generated by functional conflict can be genuinely helpful if they can show how the actual considerations adduced in opposition to a potential way of solving the conflict fall foul of one or more of the conditions on ideal conversation. That might be achieved by exposing the fact that particular kinds of desires are at odds with some of the functions endorsed in the current state of ethical practice (one identifies these functions, showing how existing precepts discharging those functions would prohibit the action or state of affairs desired); or it might be done by showing the desire is undermined by facts about the world (people who express a desire for a particular outcome do not recognize that it has consequences they strongly detest; people have the desire because of some background false belief); or it might be done by showing the desire persists because of a failure to take into account the wishes of some group of people—most obviously by showing how these people are systematically ignored by those who have the desire, but also by questioning the ways in which the balance among the variety of human desires has been struck. Familiar skeptical complaints about discussions of values typically suppose there is *nothing* for conversation to do when different priorities are assigned to the ethical functions that conflict: spoken or written words can only be the expression of an attitude others are free to reject. By contrast, there is *plenty* to be said, much that can be done to expose factual errors and false presuppositions, disharmonies with background features of the prevalent ethical code, and, most importantly, shortcomings in accommodating the wishes of classes of other people—failures of mutual engagement.

So far, I have proposed that the appropriate contemporary analog of the small band with which the first ventures in ethical practice were concerned is the entire human population, conceived as including future generations. That proposal is motivated by taking the relevant criterion to be causal interconnection: the interactions among people underlie the problem background to which the ethical project—a social technology—responds. Our forebears lived in separated groups, and the causal processes within those groups generated the difficulties to be overcome; we are affected by a far larger collection of individuals and events—contemporary human lives are thoroughly intertwined with one another.[13] On this basis, adequate ethical discussion should reflect an ideal conversation embodying the range of perspectives found in the inclusive human population, filtered by the epistemic conditions, with desires sympathetically expanded through primitive and extended mirroring. Residual divergences are to be addressed through scrutiny of the ways of balancing, embodied in rival approaches to extended mirroring. Although we cannot hope to live up to these conditions of mutual engagement, they are useful in indicating *directions* in which *actual* conversations about values might proceed (as the next section will attempt to show).

The third part of my package of proposals is an ideal toward which our ethical practices should aim: the goal of providing, for the entire population, equal and serious chances for a worthwhile life. This, too, is the scaling up of an aim that occupied the ethical pioneers (it is manifest in the ethical practices of those small groups whose way of life is closest to that of our ancestors; large lapses from achieving that goal are the particular altruism failures whose presence causes social tension and unease). Not only is the *scope* of the egalitarian drive greatly expanded (from the small band to the human population), but the *character* of what is sought embraces far more. Because the evolution of the ethical project has equipped contemporary people with a far richer collection of desires and aspirations, it is impossible for us to be content with a notion of a "worthwhile life" that satisfies only the most basic human needs (food, shelter, health, protection, family relationships)—and this remains true, even for those who struggle to satisfy the most elementary wants.

What substance can be given to the notion of a "worthwhile life"? The question has—rightly—been at the heart of Western philosophy since its beginnings, although it suffered a lengthy eclipse through a period in which religious orthodoxy promulgated an answer that invoked supernatural beings and identified communion with them as the *summum bonum*. I shall not take up the difficult task of offering an extensive answer, but it is important to fix some principal contours.

Lives go well not simply in virtue of consisting in a succession of pleasurable experiences (although the puritanical view that pleasure is always unimportant should be resisted). Worthwhile lives have a structure: those who enjoy them have a sense of who they are and what they most centrally want. Part of living well consists in having a project for yourself; another part lies in carrying out that project with some success. It is important, however, that your project not be foisted on you—it should be "your own." What does this mean? I suggest that you must enjoy the opportunity to consider a variety of options, that you are able to develop a sense of your own talents and proclivities, and that you can find among the options something that you can see as offering you (in light of the traits you recognize in yourself) prospects of realizing the central goals of the project envisaged. The possibility of formulating one's own conception of what matters most is one component of a good life—subject, of course, to the proviso that one's choice would not interfere with the like choices of others.[14] The environment in which you frame your choices must be open to real possibilities for you: it is crucial that your options not be framed in terms of arbitrary commands, effectively ruling out the most attractive. You must not be restricted by some prevailing view that particular ways of living are wicked or sinful, where these evaluations have their basis entirely in the alleged deliverances of a supernatural being.

Of particular importance is the possibility of appreciating various types of human relationships. One particular way in which some religions have distorted our conception of the human good is by emphasizing possibilities of valuable lives detached from other people. Solitary communion with transcendent beings (or with the universe), exemplified by hermits who live in remote places or those who pledge themselves to silence, is viewed as one way of living a good life. A similar thought can survive in a thoroughly secular framework: some of the most militant opponents of the world's religions commend participation in the project of understanding the natural world as an especially valuable way to live.[15] Although there is an important insight here, it needs to be carefully understood. The great discoverers achieve two things: they enjoy private states of recognizing hitherto uncomprehended aspects of nature, and they facilitate the understanding of others. The states of understanding are not the primary determinant of the value of the discoverer's life. Instead, value accrues through the contribution to understanding on the part of others.

I suggest that relations with others are central to the good life. People whose lives go well typically, perhaps always, can view themselves as having made a difference to the lives of others. Refined theoretical contemplation has its place among the catalog of factors that promote the good *life* precisely because of its potential to promote the value of good *lives*. Consequently, the life of the priest or scientist, the doctor or nurse, the teacher or social organizer, the tireless participant in the maintenance of community and family, become valuable in similar ways, through the various human relationships the person's actions sustain. The emphasis on individual freedom, on the ability of each of us to choose our own conception of what matters, needs to be accompanied with recognition that any choice that does not incorporate interactions with others and see their good as involved with one's own is highly likely to be inadequate.[16]

This judgment recapitulates an important part of the history of the ethical project. That project enabled forms of repeated interaction, through which attuning oneself to others became something to be valued and cherished. It produced ideals of human relationships and desires to contribute to enterprises involving others. Yet the evolving forms of human life do not always make it easier to achieve the valuable relationships, marked by a sense of mutual attunement and participation in a rich variety of joint projects, that are principal constituents of the goodness of lives. As human societies grow larger, even as they proliferate possibilities for living well, they may make it increasingly difficult to actualize those possibilities and to live lives with real value. In becoming coordinated parts of larger social machines, it is easy for us to think what we achieve is determined by the specific contributions we make, the deals we bring off, the things we discover or invent or compose, the tasks of whatever kind we complete, without any reflection on the impact on other people. Our self-conceptions are further debased when we measure our worth in terms of the proxies for success in any of these particular directions, the cash rewards we receive for doing them, and the trophies we thereby amass. We miss the fact that all this effort obtains its significance from effects on, or more exactly contributions to, the lives of others. We miss also the important point that, independent of any large-scale public success, lives may be interlocked in mutual dependence and mutual contribution, and thus be genuinely and completely worthwhile.

In sketching these few contours of the worthwhile life, I hope to forestall three types of major errors, to which talk of "living well" is easily vulnerable. The hedonist mistake is to decompose our lives into sequences of momentary experiences and to measure value by the balance of pleasures and pains. The individualist mistake, prominent in some religious traditions but also retained in some versions of secularism, proposes that some particular nonsocial condition of the individual—the receipt of divine grace, the making of great discoveries, the amassing of wealth—is the major source of value. The elitist mistake, already evident in the ancients' restriction of the question to the male aristocrats of the *polis*, is to suppose that something very large and uncommon is a precondition of a life's going well. By contrast, according to the view just sketched, good lives are in principle available to almost all members of our species.[17] Talk of "life projects" encourages the idea that these must be Very Grand, as if the good life required a type of distinction available only to a select few. My schematic account of the good life celebrates the ordinary. Although in almost all places at almost all times, people have been coerced or led into lives that should not be counted as worthwhile, what they have lacked are certain basic forms of freedom, everyday awareness of possibilities, not exceptional resources or unusual talents. Moreover, in many times and places, ordinary people whose lives are permeated by actions with and for others have sometimes, if not often, lived well.

I trust it is now apparent how the three proposals form a coherent package. At the center of them is a focus on the original function of the ethical project: the remedying of altruism failures. Renewing the ethical project by re-emphasizing that function can be viewed as recognizing that the problem has hardly gone away. Across the human population, we are divided by colossal failure to respond to the desires of others—for most of us, possibly even for all, many people and their aspirations are simply invisible—and the situation is compounded by variant ethical codes that announce themselves as embodying the divine will. The inclusive proposal for replacing the local band by the human species rests not only on seeing the vast web of panhuman interactions but also on regarding the ruptures and conflicts within the global population as contemporary analogs of the breakdowns that beset hominid (and chimpanzee) societies. The cognitive and affective conditions of mutual engagement are inspired by consideration of what needs to be done if altruism failures are to be overcome and sympathies expanded. Finally, in making an appreciation of human relationships central to free choice of life projects, my proposal draws on the consequences of remedying altruism failures, on the sense of a value to be found in attuning one's intentions and actions to those of others and in contributing to joint enterprises, a sense made possible by the evolution of the ethical project.

There are, to repeat, no ethical experts, only the authority of the conversation. Philosophy's role is simply one of making proposals that might facilitate the conversation. My package of proposals surely has rivals—and readers who are suspicious of that package are encouraged to formulate alternatives. Relatively friendly alternatives might allow the central focus on the original function of the ethical project—remedying altruism failures—but articulate that theme in different ways. Although there may well be superior versions among them, I suspect all members of this general family will be able to concur in many of the points of subsequent chapters, conclusions drawn on the basis of the framework outlined here. More radical suggestions, those strongly downplaying, or even resisting, the thought that the original function of the ethical project remains primary, face a different challenge, for I do not see how they can be developed as a coherent package, one that has a serious claim to conversational attention (Kitcher 2011a, sec. 56).

8. VALUES IN SCIENCE

Outlining an approach to value-judgments and discussions of values was motivated by the verdict of §4 that the value-free ideal cannot be met. I shall close this chapter by considering briefly how the framework presented can distinguish among cases in which values pervade scientific decisions.

Start with the historical examples, "scientific revolutions," in which a protracted dispute centers around the abilities of rival perspectives to solve different problems. In the astronomical transition of the seventeenth century, in the chemical revolution of the late eighteenth century, and in the acceptance of evolution (descent with modification) in the late nineteenth century, there were early phases in which it was not obvious how to assess the mix of successes and challenges. Copernicans could explain the limited separation of the inferior planets (Mercury and Venus) from the sun, but they faced difficulties in accounting for the motions of entities not firmly attached to the earth's surface (why is it that the birds and the clouds do not get left behind?) and in violating the traditional distinction between the "sublunary" sphere and the traditionally incorruptible heavens. Early in the debates between Lavoisier and his phlogistonian opponents, there were different species of reactions that each side could apparently explain. Among Darwin's initial successes were compelling accounts of the relationships among types of organisms and of the geographical distribution of species, but Darwin faced difficulties in explaining similar phenomena (electric fish, the presence of plants on remote islands), as well as puzzles about the emergence of complex structures and the notorious gaps in the fossil record.

Early in these debates, it was perfectly reasonable for the contending parties to adopt probative schemes of values, according to which their favored approach resolved more of the *important* problems than its rival. Copernicans were much taken with the "harmony" of their system; Ptolemaic astronomers with the obvious physical difficulties of a moving earth; phlogistonians supposed themselves to have adequate accounts of combustion and of the synthesis of water; Lavoisier was proud of a different resolution of combustion and a general view of acid-metal reactions; for Darwin's allies, the problems of affinities among organisms and of biogeography were critical, questions about the emergence of complexity susceptible to postponement. During the course of the debates, the constellation of problems and solutions changed. The heavens turned out not to be immutable, and everyday observations about motion undermined the worries about birds' valiant efforts to keep up with a moving earth; phlogistonian analyses of combustion proved inadequate to cope with all the experimental results; Darwinians could multiply the successes in biogeography and in understanding the relationships among organisms—even coping with the troublesome electric fish. Novel successes forced the opposition to amend ideas about what the critical issues were, to adjust the probative scheme of values. Eventually, the track record of problem solving amassed by the victors was impressive enough to deprive their opponents of any compelling probative scheme of values to support their resistance.

What occurs can be illuminated with a homely analogy. Imagine you are shopping for a used car, to be owned jointly with a friend. You and your friend visit a dealer and pick out two different vehicles. Each car has attractive features, but you both recognize that questions remain unresolved. As you continue to run various checks, your friend's choice passes with flying colors: the engine is clean and well maintained, the transmission is smooth, the interior has been carefully kept up. By contrast, your own initial preference turns out to have a dubious service record, the engine makes odd noises at particular speeds, gear shifting is erratic, several parts of the dashboard are loose, and the tires are threadbare. You would feel little temptation to insist on your original choice when all you can point to is an extra pouch for containing documents and an appealing hood ornament. Taking those features to confer higher value would be absurd.

Matters are similar in the historical instances. Debates reach a stage at which no serious scheme of *cognitive* values would yield a scheme of probative values capable of prolonging debate. If some people continue to resist, they do so because of allegiance to *broad* schemes of values that shape judgments about cognitive and probative values. Persistent antagonism to evolutionary theory provides an obvious illustration.

For all the official rhetoric of the anti-Darwin movements, it is evident that the people drawn to it lack any well-articulated scheme of cognitive values, in which a host of biogeographical explanations are dismissed as peripheral and the provision of a detailed scenario for the emergence of the first cell is taken to be an absolutely crucial problem, one sufficient to outweigh any amount of success in other areas. Maintaining a view of the history of life consistent with a particular reading of the scriptures overrides other considerations. Even if the value-free ideal were abandoned, it would be possible to diagnose the errors of the Darwin doubters by amending the ideal: probative and cognitive schemes of values could be allowed into scientific practice, broad schemes debarred. I recommend a different diagnosis. The error lies not in invoking broad schemes of values, but in *the character of the particular broad scheme of values introduced.*

Here the framework of the last three sections comes into play. Broad schemes of values can play a legitimate part in scientific practice, but they are required to be sustainable in an ideal conversation. It must be possible for a discussion, under conditions of mutual engagement, to endorse the values in question. The anti-Darwinians fail because the proposals on which they rely cannot be part of a conversation of this sort—most evidently and straightforwardly because they violate the cognitive conditions on mutual engagement.[18]

To see why this is a *better* diagnosis, turn to the family of examples in which researchers make serial decisions in a line of investigation. They tacitly appeal to probative schemes of values, without broader reflection. Most of the time, perhaps virtually all the time, an ideal discussion, under conditions of mutual engagement, would endorse those schemes of values and their application to the decision at hand. Any consequences for broad values are at best only dimly foreseeable; there are no grounds for unusual caution or for proceeding with particular urgency, and thus no basis for accusations of recklessness or inappropriate irresolution. Sometimes, however, the decisions do have consequences a well-supported broad scheme of values would view as important: there is an opportunity to develop a drug that might bring great benefit to people, or risks of generating a situation producing considerable harm. Under these circumstances, we would expect the scientist to adjust a probative scheme of values that might be entirely adequate in the normal instances. Not only would an ideal discussion condemn decisions made using an unendorsable broad scheme of values (as in the example of the anti-Darwinians) *but it also would frown on failure to deploy an endorsable broad scheme of values*: a scientist who stubbornly continued with an original plan for research, when an unanticipated discovery made along the way opened up new opportunities for relieving human suffering, would reasonably be criticized. The value-free ideal breaks down because of the importance of using well-grounded broad schemes of values (schemes that would be approved by an ideal discussion under conditions of mutual engagement) to govern a program of research.

Return finally to the imagined frank climate scientist. The scientist's recommendations rest on the assessment of some mix of potential outcomes as sufficiently terrible and sufficiently likely to require action to prevent them. We can imagine detailed explanations both of why some of the scenarios are genuine threats and of the values they would endanger. In the end, our interlocutor will ask us to agree in the badness of particular forms of human deprivation and suffering. That request is entirely reasonable, for the value-judgments deployed are among the most straightforward conclusions any ideal discussion under conditions of mutual engagement would reach.

By contrast, an equally frank *opponent* of environmental action, who confessed to the overriding importance of maintaining the profits of a particular firm, would reasonably be seen as failing to live up to the proper standards of scientific advocacy—not because of the mere entanglement of values, but because the specific value-judgment presupposed would be dismissed by an ideal discussion under conditions of mutual engagement. There is an important difference between the justified appeal to values by the responsible climate scientist and the values invoked by industry partisan, a difference brought out by the framework offered in previous sections.

That does not entail, however, that the only legitimate positions in this debate are those favoring climate change activism. *Responsible* opposition to the views of the frank climate scientist would not rest on the partisan values but on judgments that could be endorsed in ideal discussion. Indeed, the most obvious species of opposition would endorse the *very same* broad scheme of values as that underlying the call to environmental action. Serious challenges acquiesce in the badness of large-scale human deprivation and suffering, arguing that the interventionist measures are more likely to increase these unfortunate consequences by hampering the accumulation of resources future generations will probably need to solve their problems: it's not simply that people will have less (real) money, but that the quality of their lives will be diminished.[19] Ironically, in this most pressing of contemporary policy debates, the approach to assessing values outlined here offers a relatively easy way to generate a shared basis for debates about what to do. As we shall see later on (§41), the principal issues concern the balancing of sacrifices that can be recognized from a scheme of values disputants hold in common.

Chapter 3

DEMOCRATIC VALUES

9. TAKING DEMOCRACY SERIOUSLY

The central task of this book is to offer a theory of the place Science should occupy in a democratic society, a theory about how the system of public knowledge, in which Science is so prominent, should be shaped to promote democratic ideals (§3). To develop any such theory, a picture of democracy is required. This chapter will build on the approach to values just outlined, to elaborate a view of democracy and of democratic ideals, *in which a healthy system of public knowledge is seen as essential.* The view will surely not correspond to most people's first thoughts when they reflect on the character of democracy,[1] so it is best to start with the apparently simplest conception.

Many popular discussions of democracy—for example, in contexts where politicians advertise the spread of democracy to areas they have invaded—take the existence of elections as a talisman. Not just any elections, of course, for it is important that citizens of a democracy not be coerced into voting for some thug who has gained power, and that they be able to discuss and debate the issues openly and freely. Given these background conditions, however, a common definition of a democratic society would take it to be one in which all the mature citizens have the opportunity to vote for the leaders and representatives who will make the political decisions.

Despite its currency, the voting conception is quite inadequate as an answer to the question "What is democracy?" Part of the inadequacy stems from the fact that a good answer ought to bring out the reasons why democracy seems valuable, why we might want this kind of society. As Dewey noted, questions about the health of democracy ought to concern thoughtful citizens, and, when we try to address such questions, we need to understand the values democracy is taken to promote: it is insufficient simply to think about "the ballot box and majority rule," for these are merely "mechanical symbols" (1958, 36; 1985, 144).

The external manifestations express an ideal of popular control. Voting provides citizens with an opportunity for input into the decisions that affect their lives. As Ian Shapiro puts it: "Democrats hold that governments are legitimate when those who are affected by decisions play an appropriate role in making them and when there are meaningful opportunities to oppose the government of the day, replacing it with an alternative" (2003, 5). Although this is vague—"appropriate role" and "meaningful opportunities" can cover a lot of territory—Shapiro's formulation points toward something important. Similarly, Robert Goodin's concise statement of his own answer brings out a crucial aspect: "Democracy is a much-contested concept. Fundamentally, though, it is a matter of making social outcomes systematically responsive to the settled preferences of all affected parties" (2003, 1). Here the significant addition is the introduction of the notion of a "settled preference"—roughly something opposed to an uninformed or rash impulse—a notion rightly given prominence in Goodin's treatment of democracy.

The emphasis on voting and popular elections derives from a sense that this is the best way for the people to exercise control of those who make decisions on their behalf.[2] Yet there are good reasons to wonder whether the counting of votes and the actualization of the majority view serves to express the "will of the people" or even anything valuable. In his probing analysis of the abstract conditions for democracy, Robert Dahl introduces an important idea, the thought that one's individual viewpoint need not be adequately captured in a vote for the least bad alternative among a set of dreadful choices. Dahl suggests an ideal condition that ought to be met if the citizens play an "appropriate role" (in Shapiro's vague phrase) in decision making: "Any member who perceives a set of alternatives, at least one of which he regards as preferable to any of the alternatives presently scheduled, can insert his preferred alternative(s) among those scheduled for voting" (1963, 70).[3] There are surely instances in which the failure to meet this condition undermines the value of a majority vote, as in Dahl's own example, when the choice is between voting for the Supreme Leader and not doing so (and going to the gulag).

Dahl raises another difficulty for a bare reliance on majority vote. As he recognizes, it is quite possible for there to be a disagreement on an issue where one group of people are relatively indifferent in their view, while their opponents care passionately about how it is resolved. So there can be a majority in favor of a position, comprising people who would not mind much if the vote went against them, and a slightly smaller minority who feel deeply affected by the outcome. If this supposition appears to presuppose an allegedly impossible comparison of interpersonal preferences, it is worth noting that the intensity with which an outcome is valued can often be assessed (roughly, to be sure) by considering the ways in which individuals are prepared to sacrifice time and comfort to their causes; if you think that the preferences of people who are prepared to stand out in soaking rains, to spend hours on long marches, even to risk beatings and torture are not comparable to those of others who would change their minds on the issue if it would cause them a slight delay in having their next meal, you have to believe that human valuations of simple sensory experiences (being cold and wet, being numbingly fatigued, being beaten and brutalized) are strictly incomparable, even when those involved have no apparent physiological or psychological peculiarities; that is surely an extraordinary supposition, a relic of operationalist dogma.[4] If so, the scenario of the relatively indifferent majority and the passionate minority makes sense. In such instances, the majoritarian approach to voting fails as an adequate expression of popular sentiment, for "popular sentiment" would be better measured by the total amount of *feeling* directed toward one option rather than by the number of *people* who incline one way rather than the other.

The existence of elections and of majority rule is not constitutive of democracy. Often, these serve as the expression of a deeper idea, that of popular control. Nevertheless, they may not even be *expressions* of that idea but *betrayals* of it. We have to dig further, looking for an account of popular control and why we should want it. Everyday rhetoric about democracy provides direction: democracy is the key to political freedom (recall Plato's judgment about what the friends of democracy would say [§2]). This is only half the story. The appeal of democracy rests on two ideals, one of freedom, and one of equality. But the best place to start will be with versions of the ideal of freedom.

10. IDEALS OF FREEDOM

Isaiah Berlin's (1961) celebrated inaugural lecture crystallized centuries of thought about political freedom. Berlin distinguished two main traditions:

one, the *negative* conception, treats freedom in relation to threats—freedom is "freedom from"; an alternative, the *positive* conception, looks toward the possibilities that freedom opens up for us—freedom is "freedom to." Although the division is valuable, it should be further articulated in two ways. First, there are important variants within each of the two main types; second, it is important to detach the notion of positive freedom from a totalitarian conception of the state (with which it is frequently associated).

The idea of negative liberty has been prominent in the Anglo-American political tradition. It goes back to Hobbes, who famously frames his political theory by offering an account of a "state of nature" in which individuals are inclined to interfere with and block the projects and aspirations of others. This interference from others is the antithesis of freedom, and the transition to a state makes us free to the extent that it diminishes or, ideally, eliminates, such interference. There is, of course, an obvious danger, made vivid in Hobbes's own proposal of an absolutist state, a great Leviathan. The existence of the all-powerful sovereign may prevent other individuals who would otherwise have foiled my endeavors from interfering with me, but the state itself may generate new modes of interference. Indeed, if human motivations are as Hobbes thinks they are, and the inclination to interfere with others so entrenched, it is hard to see how the state could *not* produce new types of interference, for *ex hypothesi* at least some of us (and probably all of us) have the tendency to interfere with others, and if the state is to offer its needed protections, it will have to thwart these tendencies—and the thwarting will itself be a sort of interference. There are losses, as well as gains, in the transition to the Hobbesian state, and an adequate defense will have to take the form of showing that, on balance, we come out ahead.

This is not an artifact of Hobbes's particular conception of how the state is to be run, a consequence of his absolutism. The point applies generally. If there is a population of people whose aims conflict, and if each is bent on pursuing his/her aims, then, without regulation, however decided, there will be some form of interference; equally, *with* regulation, however decided, there will be interference. Hence, with or without a state, there are patterns of interference—different patterns, to be sure—and if the state can be said to promote freedom, then it must be that the kinds of interference it produces are more tolerable (less invasive of freedom) than those it eliminates.

The negative conception can be developed in a number of ways from this point. The starkest possibility would be to declare that all interference is bad, and the less of it the better, so the task of the political state is to minimize the amount of interference (the crudest way to do this would be simply to aggregate, but more nuanced approaches are plainly possible). Another option would be to declare that there are certain natural rights, shared by all human beings, and that states provide freedom by eliminating those forms of interference that would violate those rights. On this basis, one could compare the distribution of interferences produced in different political states and declare that one offered greater freedom than others because the interferences it generated were directed at preventing violations of natural right, rather than blocking actions that are neutral to questions of natural right or even introducing violations of natural right. An attraction of the approach is that it avoids any problem of how to aggregate interferences, and it also allows the plausible conclusion that not all forms of interference are bad. Rather, we should welcome those that protect people against violations of their rights. (When Punch wants to hit Judy over the head, it is a good thing the officers of the law interfere.)

Given the perspective on the ethical project adopted in the last chapter, the idea of "natural rights" existing independently of human deliberations and awaiting discovery by insightful political theorists is quite hopeless. Attribution of rights depends on what limits would be set in progressive elaborations of our social technology, what functions would be endorsed in conversations under conditions of mutual engagement. *The idea of negative freedom, conceived in terms of protections of rights, already presupposes a deep democratic concept, that of discussion on terms of equality.*

A similar point emerges if we consider a different tradition of developing the negative conception of political liberty. Those, like Bentham, J. S. Mill, and Marx, who think vague appeals to rights can provide cover for social injustice, seek an alternative foundation. Mill's classic formulation proposes that individual lives are centered on projects, embodying people's conception of what matters to them and who they are, and that political freedom consists in not interfering with those projects except insofar as they intrude on the projects of others. A free society contains a set of private spaces, one for each of the citizens, so that there is no interference with a citizen's operations within his/her private space. There is an important distinction between those forms of interference that reach into a private space and those that protect the private spaces; the former are breaches of freedom, while the latter are constitutive of it.

Any approach of this kind must draw the boundaries of these private spaces, and Mill and his successors have invoked the notion of harm in their attempts to fix the limits. Only those of your actions that *harm* others are proper targets of outside interference. Many of the things people do cause difficulties for others in the pursuit of their individual projects. Which of these constitute *harms*? The history of liberal political thought after Mill demonstrates the difficulty of the question. As with the conception of a prior set of "natural rights," the thought of "objective harms," awaiting recognition by insightful theorists, is belied by reflection on the character of the ethical project. Once again, the boundaries are to be drawn in our elaboration of our social technology, fixed through conversation under conditions of mutual engagement.[5]

Turn now to the positive conception of freedom. Some writers (for example, Rousseau) understand the notion of freedom in terms of self-mastery. Two fundamental ideas underlie this approach: first, people are conceived as having deep desires, even needs for joint activity with others; second, all of us have desires and passions that can easily lead us in incompatible directions. Our social nature inclines us to participate with others on cooperative projects that will benefit all those who join in. On other occasions, we may feel a strong urge to desert any such project to pursue some goal that appeals to us personally or that will bring a beneficial outcome for some smaller group (our family, our intimate friends). The crux of the social contract is that the parties to it first recognize the entire group as one in which the potential for fruitful cooperation exists and then see the desirability of such cooperation.[6] At the moment of contract, they commit themselves to a constant policy of always preferring the socially good outcomes, those that would bring the benefits of cooperation to all, to results that might conform more closely to their erstwhile, individual predilections.[7] If, later on, someone is tempted to deviate, then, when the others bring him into line, he is really "forced to be free," in the sense that his temporary, personal preferences are overridden in the interests of the socially directed outcome to which he committed himself at the moment of contract and which express his deepest—social—nature.

The conception of freedom developed here supposes that the imposition of the law liberates us through recalling us to the deepest—social—aspirations we have. In an important sense, we fashion ourselves through commitment to sharing a society with others because we create an order that governs our competing tendencies. Thinking about freedom in this way embodies one of the ethical themes of the last chapter; to wit, that joint action with others is central to any satisfying personal project. Yet it is evident that this positive conception can be perverted, as when the parties to the social contract are manipulated into committing themselves to some worthless, or even oppressive, "cooperative endeavor," the glorification of the Leader or the triumph of the Fatherland. To block these difficulties, we should require that any genuine form of cooperation must be recognizable as bringing real benefits to all—where the assessment of benefits is grounded in ideal discussion under conditions of mutual engagement. Moreover, the requirement must be interpreted to include all those affected by the activity, even people who are not numbered among the "cooperators." If the parties to one social contract would all benefit from a certain kind of outcome that would threaten the parties to a different contract, the failure to achieve mutual benefit on a broader scale undermines the identification of their joint activity as a form of valuable cooperation. Conceiving freedom as self-mastery, involves, in its democracy-friendly variant, a constraint from the side of negative liberty: the capacity for self-mastery instituted within any group is limited by the requirement that the like capacities of other groups be respected. Positive *democratic* freedom presupposes the inclusive conception of the community of deliberators, defended in §7.

Dewey offers a related positive conception by starting from the Millian picture of the free society with its protected spheres of individual self-realization. He sees this as self-realization in name only, for the significant projects that attract people, those around which they would, after reflection, want to center their lives, involve interactions with others. A guarantee of your own personal protected space—and no more—provides an opportunity for only a stunted life. The projects that really matter are not projects-for-me or projects-for-you but projects-for-us. Self-realization consists in the capacity for undertaking these, and genuine freedom occurs when that capacity becomes available.

In his attempts to probe the notion of democracy beyond the superficial phenomena of voting and elections, Dewey suggests that "[a] democracy is more than a form of government; it is primarily a mode of associated living, of conjoint communicated experience" (1997, 87). Crucial to the achievement of genuine democracy is a process of breaking down the barriers, not only between individuals, but also between larger groups, between classes, races, and nations. "These more numerous and more varied points of contact denote a greater diversity of stimuli to which an individual has to respond; they consequently put a premium on variation in his action. They secure a liberation of powers which remain suppressed as long as the incitations to action are partial, as they must be in a group which in its exclusiveness shuts out many interests" (Dewey 1997, 87). The same theme recurs in an essay written twenty years later: "The keynote of a democracy as a way of life may be expressed, it seems to me, as the necessity for the participation of every mature human being in formation of the values that regulate the living of men together: which is necessary from the standpoint of both the general social welfare and the full development of human beings as individuals" (Dewey 1958, 58). These, and many similar passages in Dewey's writings, offer both an ideal of freedom and a claim about the relation of freedom to democracy. The ideal proposes that freedom consists in a certain sort of self-realization; to wit, the development of the human capacity for participation in joint cooperative activity with others. This sort of freedom is a crucial component in a truly democratic society.[8]

The resultant conception of free society is more complicated than that proposed by Mill, because it envisages a large number of levels at which protected spaces are needed. There are families, groups of friends, teams, communities, unions, professional societies, social movements, churches, political parties, and nations—and the species as a whole—all of which engage in joint projects. Individual people typically realize themselves not just by engaging in projects they carry out in isolation, but also by being a part of many of these larger groups, and the division of individual effort among agents of different kinds is typically idiosyncratic. The Millian conception of "finding one's own good in one's own way" should thus be taken to require a kind of positive freedom, the freedom to act in community with others, to realize oneself through joint participation in projects. At the same time, that positive freedom depends on an extension of the kinds of protections Mill recognized, in that joint activities at all levels have to be protected except when they intrude on the like activity of others.[9]

Given the common themes that run through them, the four approaches to freedom just reviewed need not be seen as competitors. They offer different ways of elaborating the central ideas about values, and their assessment offered in the previous chapter, celebrating the central importance of individual projects, of social interactions in worthwhile individual projects, and of decisions reached under conditions of mutual engagement. We do not have to decide on *one* of these ideals and adapt our concept of democracy to it. We can take them all seriously, bearing their different emphases in mind as we seek to understand and improve the democratic predicament in which we find ourselves.

It is worth being explicit, however, about the difference between *all* these fundamental conceptions and the crude and superficial ideas disseminated by many people who declare themselves to be fans of political liberty. To the extent that the protection of private property is an expression of freedom, it is because the ability to use particular resources as one sees fit is sometimes important for promoting worthwhile life projects. Any connection with keeping large amounts of one's earnings and resisting taxation is far more tenuous. Without public revenues, goods essential for undertaking joint projects are not available, so the fundamental freedoms of all are diminished. Strident attacks on measures for generating those revenues are often the most misguided—or cynical—debasement of the political value they claim to cherish.

11. EQUALITY IN FREEDOM

Of the many respects in which people might be held to be equal or unequal, several are pertinent to consideration of democracy. No democrat should be committed to the doctrine that all people are equal in their talents and abilities, although one might hold that the social manufacture or amplification of inequalities in performance of various kinds represents a failure of the democratic ideal. Similarly, one might focus on inequalities in distribution of resources, arguing that democracy requires justification for any deviation from equality in this regard (Rawls 1971). I shall start with a relatively thin conception of equality and approach more ambitious possibilities in steps.

The barest conception of democratic equality focuses on the procedures through which the society is run, the surface manifestations of democracy. Citizens in a democracy may be thought of as equals insofar as they are treated equally with respect to these procedures, each being allowed to and required to do the same things. Just as we cannot understand democracy without considering what the procedures are *for*, so, too, we cannot identify democratic equality unless we see why this formal equality with respect to procedures is a good thing, how it serves important purposes. Although we have not yet probed the ways in which the procedures associated with democratic political arrangements might promote the kinds of freedom just considered, it is helpful to envisage the ideal of equality to be sustained by these ideals of freedom. We value the procedures of existing democracies because they promote the freedom of the citizens; we want there to be formal equality, equality with respect to these procedures, because we want a certain kind of equality with respect to the freedoms enjoyed; hence the ideal of equality that underlies democracy is some sort of equality with respect to these freedoms.

This is vague, but it serves to narrow the inquiry into those respects in which equality is required by democracy. With respect to the ideals of the previous section, should the pertinent concept of democratic equality be understood as equal *achievement* of those ideals or as some type of equal *access to*, or *opportunity for*, or *capacity for* those ideals? Start with the conception of negative liberty.

Whether we approach the issues in terms of rights or of protected spaces, the state is conceived as setting up mechanisms that protect against a certain type of interference, and, although it does so by creating a different form of interference, this latter form is taken to be relatively benign. The machinery of the law is introduced to secure natural rights or to guard the private sphere in which individuals pursue their projects, and the costs of having that machinery, the consequent restrictions they place on citizens, are taken to be worth paying in order to obtain the valuable ends. A very obvious way to construe the democratic ideal of equality is to suggest that, in the creation of the mechanisms of law, all citizens are treated equally. Each receives the same protections, and each is required to abstain from the same kinds of intrusions into the lives of others. It is an important part of this ideal of equality that the protections are genuinely afforded to all, that some citizens are not more exposed than others to violations of their rights or invasions of their individual projects. Nor can the burdens of supporting the institution of law be distributed unevenly, so the effective costs for some are much higher than the effective costs for others. To require all citizens to perform a particular type of action may create a substantial inequality if there are some for whom the action is readily done and involves no lessening of the ability to pursue private projects, while for others the performance is incompatible with any satisfying way of life. Because burdens usually accompany the actions required to sustain protective laws, the idea of equality before the law, conceived as equality in respect of negative freedom, is typically incompatible with large inequalities in resources. Appraising the democratic condition of a society with respect to the negative ideal of freedom effectively involves posing two kinds of questions: questions about the kinds of interferences prohibited by the law and the kinds of interferences produced by the law, on the one hand, and questions about the distribution of protections and burdens across the citizenry, on the other.

Positive conceptions of freedom generate more demanding kinds of equality. Suppose freedom requires the ability to recognize, and effectively choose among, a wide array of possible life projects (whether viewed as strictly individual ventures or programs of joint activity). To join this with an ideal of equality in freedom would be to make fundamental a certain type of equality of opportunity, not understood in terms of the equal possibilities for acquiring resources but as the equal availability of a rich menu of options for one's life—*just the conception of the good that figured in §7.* It would surely be unrealistic to demand that exactly the same (rich) menu be available to each of the citizens, for, once differences in individual traits are recognized, that is impossible: the tone-deaf are never going to make it as opera singers. The ideal does, however, require certain choices: we should prefer to expand the range of options for those whose menus are currently more limited, even though that might mean foregoing expansion for the fortunate in a more impressive way; if differences are very great, diminution of the options for the fortunate can even be required to promote more choice for those whose prospects are very limited (Rawls 1971; Mill 1970).[10]

The Rousseauian positive conception of freedom centers on the commitment of participants in the social contract to the common good, and this notion of the common good has an ideal of equality built into it. For a society to be free in Rousseau's sense, the joint actions of the citizens must be directed to the common good, and that is incompatible with the continued increase in the inequality of the distribution of returns. Hence in this case, the ideal of equality is not an addition to the ideal of freedom but a consequence of it.

I have outlined general reasons for thinking that the distribution of resources in a society is likely to be related to the distribution of freedom within it. Considerations of this kind are secondary to detailed examination of the ways in which particular societies succeed or fail to live up to their professed democratic ideals. Yet they should motivate political theorists—and concerned citizens—to undertake that examination. One of the great scandals of the clamorous insistence on "individual freedom" in supposed paradigms of democracy is its profound neglect of issues about the distribution of any form of freedom worthy of the name. Many of the most vociferous champions of "freedom" have only *their own* "freedom" in mind, and the things they want done would further diminish the freedom of people whose real liberty is already far less than their own.

12. A PICTURE OF DEMOCRACY

We started with democracy and with its surface manifestations, votes and elections. That led to vague, though suggestive, phrases about people exercising joint control over decision making. Behind those phrases I discerned ideals of freedom and of equality. It is now time to try to fit these pieces together.

The basic form of the picture is obvious. We have three levels.

1. Ideals of freedom and equality in freedom
promoted by
2. Involvement of citizens in decisions about the matters that affect them
realized in
3. The standard machinery of elections, votes, the law, etc.

Setting out the picture explicitly raises two very obvious questions: First, how, in general, do the relations among the levels work? Second, what relation does the picture have to the societies we call "democracies"? Start with the first question, specifically with the idea that involvement of citizens in decision making promotes ideals of freedom and equality.

On some conceptions of freedom and equality the relation is straightforward. If you think of freedom as self-realization, and conceive self-realization as involving the pursuit of joint projects, there is little mystery about why joint decision making, in which citizens participate as equals, promotes the ideal of freedom and equality in freedom—the activity identified as a means is itself involved in the end to be promoted. Of course, adopting this conception does raise doubts about whether the institutions found in actual democracies are proper realizations of the ideal: laws that protect all citizens equally, produced through elections at which people vote, do not seem to be enough; there must be real discussion among the participants, something more like the town-meeting democracy of eighteenth-century America.[11] If votes and elections are connected with the promotion of the underlying ideal of self-realization, it must be because they are necessary conditions for attaining that ideal, for, by themselves, they are plainly not sufficient.

So, too, for the other version of the positive conception of freedom, Rousseau's ideal of self-mastery through acquiescence in the general will. Although the link between this ideal and participation as equals in discussion and decision making is more tenuous—for, as totalitarian developments of Rousseau's approach have revealed, the voices of state authority can at least *claim* to identify the common good without popular discussion—one might plausibly hold that the citizens need to meet and discuss if they are to honor their commitment to the common good. The public forum can be seen as a place in which all citizens participate in the education of each. It is far from clear that merely recording a vote can serve the same purpose, revealing to the minority that its view was at odds with the general will.[12]

When we turn to the negative conceptions of freedom, however, the connection between the ideal of freedom and of equality in freedom and the public decision making supposed to promote it becomes questionable. The mere institution of majority rule in regular elections brings no guarantee of decreasing interference in ways that protect rights or personal projects, or of doing so in a way that treats all citizens equally, nor even of making it more probable that these outcomes will ensue. Tocqueville and Mill both recognized that the majority can impose its own form of tyranny, and Mill drew the explicit conclusion that the problem of promoting freedom had been miscast by viewing it as connected with majority rule. The link between ideal and procedures is forged differently. Even if we suppose joint decision making—citizen involvement in decisions about the matters affecting them—there is no change in the situation. To the extent that the underlying ideal of freedom and of equality in freedom is attained, it is because constitutional provisions debar certain kinds of interferences, intrusions into the private sphere or violations of rights. If public discussions, public decision making, votes, and elections play a role, they do so through setting these provisions in place and protecting them when they are there. Going to the polls may not do that; indeed, it may even have a contrary effect. There is only a loose fit between the underlying ideal and the decision making supposed to promote it.

The point should be obvious from the lack of synchrony between the extension of democratic procedures to include people who have previously been left out, and the achievement by those people of anything like these forms of negative freedom. Permitting propertyless men, women, and members of ethnic minorities to vote failed to eliminate all the distinctive ways in which interferences reached into these people's lives, nor did it secure equal rights for them. If there has been progress toward extending the ideals of negative freedom to encompass these groups, it has come about because of the *further* social efforts undertaken by some of their members (often aided by a few sympathizers). American women were awarded the vote shortly after the first World War, but it took the Women's Liberation Movement of the 1960s to increase negative freedom—and, of course, the Equal Rights Amendment has still not been passed.[13] The joint decision making, realized in elections and voting and popularly regarded as the core of democracy, plainly does not suffice for the ideals of negative freedom. Moreover, strictly speaking, it is not even a necessary condition, since the constitutional provisions sustaining the negative freedoms might be instituted and maintained by wise and benign rulers: one can imagine people with a generalized version of the attitudes of the most high-minded abolitionists.

Hence there is considerable looseness among the levels of the official story about democracy (or the official stories, since there are variations in the underlying approach to freedom). The looseness disappears from view because we naturally import into the narrative features of the historical processes through which our democratic institutions emerged. We recall a specific way in which forms of interference or of violation of rights became salient for groups with sufficient resources to try to resist the infringements of negative freedom. As we think about the historical shifts we characterize as steps toward democracy, we can recognize a pattern. Within a society, there are two groups, the oppressors and the resourceful oppressed; these groups need not exhaust the society, for there can easily be much larger populations whose members are even more intensely oppressed and have no resources for resistance. The oppression felt by the resourceful oppressed can typically be characterized in terms of actual interference or in terms of rights violations. (Rival variants of the negative conception are readily fitted to narratives of the emergence of democracy.) The resources of the resourceful oppressed are sufficient to enable them to overthrow the oppressors, or at least to limit their capacities for action, and to set in place protections against the forms of oppression previously experienced. For those protections to be secure, the subsequent administration of them must be subject to check by the members of the group to be protected. The obvious form of the check requires members of that group to be involved in a range of relevant decisions, and to have the power to vote on the administrators of the pertinent sphere. The decision making and the elections constitute a mode of control that attempts to block the danger of reversion to the previous state of oppression.

This pattern can be traced in many historic episodes: in the history of Rome (both the Republic and the Empire), in the careers of Italian city-states, in the emergence of the British parliamentary system from Magna Carta on, and in the American Revolution and its aftermath. In all these instances, the passage of control from the oppressors to the erstwhile oppressed is achieved by involving the latter in decisions of the kinds that previously led to limitations of their negative freedom. There is a *specific* problem with respect to negative freedom, the problem of the *identifiable oppressors*, to which democracy, as conceived in the official story, offers a solution. That problem arises when the invasions of negative freedom can be attributed to a source of oppression (a tyrant or a group that acts tyrannically). If that source can be overthrown, or limited, and if actions of the same type as the previous invasions are then subject to the joint control of the potential targets, the problem can be solved.

As Tocqueville and Mill saw, the new system of joint control can introduce a new version of the same problem, in that a majority in the newly enfranchised group can invade the freedom of the minority in just the ways previously found to be oppressive. Even if a constitutional provision is set in place, it is evident that the guarding of the constitution cannot be achieved through democratic decision making, voting, and majority rule—for that would only generate a new occasion for a majority to exercise its tyranny. Consequently, as constructive politicians from Madison on have seen, other means must be devised, and, as we might suspect, the institution of these means can easily conflict with some underlying ideal (particularly with respect to equality in freedom).[14]

My chief concern, however, is not with considering how the democratic response to the problem of identifiable oppressors should be articulated to avoid newly generated tyranny of the majority. Two different points are relevant to the issues of the character of democracy and the place of Science in a democratic society. First, if we are to understand the relations among the levels in the picture of democracy that has emerged, we do better to frame our questions in terms of the continued process of *democratization*, so the task is always to think about the institutions we have inherited, how they have responded to earlier difficulties in realizing the underlying ideals, and how they might be further developed and refined to promote those ideals. Second, the problem that has dominated the *past* efforts at democratization presupposes an identifiable source out of which the limitations on negative freedom flow. One way for oppression to persist, even in the presence of public decision making, elections, and the usual democratic machinery, is for the source of oppression to be very well disguised. If the voters cannot see what is happening, or if they are unable to trace the confinements they feel to their source, their votes are unlikely to serve as a means of control. The decay of functional democracies into totalitarian regimes reveals one way in which this can occur. My interest lies in a different possibility, one characteristic of the large societies of the modern world, the problem of *unidentifiable oppression*, where the limitations on freedom are either not felt, or, if felt, are difficult to trace to their source because no single agency is involved (or when the role of any human agency is indirect). *With respect to this problem, public knowledge is essential.*

We can readily imagine small societies in which the relations among levels in the three-tier picture of democracy would take the straightforward form depicted. Given a low number of decisions to be undertaken, and a small group, the decisions could be achieved through thorough discussions in which all parties were treated as equals. You can elaborate the account so that it enshrines any of the ideals of freedom and of equality in freedom I have distinguished. Existing bands of hunter-gatherers—as well as our forebears who began the ethical project—might approximate the egalitarian democratic ideals (Lee 1979; Boehm 1999). That ideal *may* also have flourished in the small New England communities, whose town meetings Tocqueville so admired.

Within contemporary industrial and postindustrial democracies, however, the idea of public control of decisions that affect all citizens looks ludicrous. An intricate division of labor means the life of any individual is affected by the actions of vast numbers of others, so enormous numbers of institutional mechanisms need to be in place to constrain interactions taking place at unfathomable distances. To set in place regulations governing all these actual and potential interactions, and to administer those regulations, requires a large number of decisions, each of which can affect the lives of many citizens. No single individual can hope to have a clear view of all the likely consequences of a new proposal, at least not until experts from many different fields have pooled their insights. Consider, for example, a suggested modification of speed limits. The modification may affect you and others in any number of ways: it may make it more or less likely you will have an accident, it may increase or decrease the time you spend traveling, it may lower or raise the levels of pollutants in the atmosphere, it may make it more or less attractive for goods to be delivered to various regions, it may encourage a shift in the mode of transporting those goods, it may require different forms of maintenance for various modes of transportation, and so on and on. You are in no position to assess all these consequences or even to recognize some of them. Nor are those charged with decision-making responsibility in any better position to do so—at least not until the various experts have contributed their pieces and the synthesizers have produced their summaries. We are affected by numerous decisions we delegate to others, and those others delegate further, until, in the end, an ungainly distributed entity reaches a conclusion.

As Robert Dahl pointed out in a classic discussion, any ideal of full participatory democracy is quite unfeasible for a large society whose members must make many decisions. Imagine some significant number of decisions arises within a year, and everyone is allowed some amount of time to say his or her piece on each. The time needed for participation and deliberation will have to be the product of the number of citizens, the number of decisions, and the amount allotted for each individual speech. A group of one thousand citizens, confronting one hundred issues a year, could allow five minutes per person per issue—assuming complete dedication to the project and universal insomnia; if the number of citizens is increased to one million, each person's allotment drops to 0.3 seconds.[15]

This means, of course, that some conceptions of democracy, based on positive ideals of freedom (self-realization through participation in decision making with others, education in the content of the general will through discussion with others), are simply impossible unless the product of the number of citizens and the number of decisions is small. Even if you suppose democracy requires only an expression of preferences with respect to the decisions the society must take, simple voting on those issues, there are still troubles. For the citizens must devote time to understanding the questions posed to them, and, on average, this will be nontrivial. If one thousand issues arise each year, and if citizens devote twenty hours a week to preparing themselves to vote, they will be able to devote one hour to each question. When you consider the range and complexity of questions about defense, health policy, employment, education, and environmental impact, even people who process information quickly will be unable to explore everything thoroughly.

Genuine participation in all the decisions that affect the lives of citizens is obviously impossible. You might think the problem could be ameliorated by focusing on those decisions that make the most serious differences to the citizens. Perhaps many of the issues are unimportant and could be struck off the agenda without great loss, leaving people free to concentrate on the remainder. A moment's reflection, however, brings home the point that there will be few, if any, social decisions that are trivial for *everyone*; it is not that there are some consequential for all, and others consequential for none, but that, for each person, there is a set of consequential issues; if each of us knew which these were, we could concentrate on the ones pertinent to ourselves and agree not to cast an ignorant vote on the rest; but of course, finding out which set matters to you requires enough understanding of the entire agenda to filter out the irrelevant ones, and we are back with the need to devote significant amounts of time to inquiry.

The way out, taken by all large democratic societies, is to pursue a system of government the ancients, who introduced the term, would not have counted as a democracy. With a few exceptions (issues requiring referenda), we delegate the work of decision making to people who represent us. They, in their turn, delegate to others both the task of obtaining concise summaries of the considerations bearing on complex issues and the work of elaborating the concrete policies emerging from their decisions (Richardson 2002). A highly distributed agent produces a detailed policy to direct actions, and those actions will combine with many other policy-directed actions, themselves produced by similarly distributed decision making, to yield consequences for the lives of the citizens. Frequently it is not just that citizens do not make the decisions but that the consequences come about without any person being able to foresee them and approve them.

How, then, do citizen-voters have control over the decisions taken in their name? The classic answer (Schumpeter 1947) is that the representatives offer the citizens *packages of positions*, subject to prospective assessment, so citizens can vote for whichever package most appeals to them. These packages can be judged retrospectively, so citizens can replace representatives whose decisions have turned out to be at odds with their wishes. But this notion of citizen control is anemic at best. Not only do the voters have no option of inserting into the package the options they would prefer (Dahl 1963), but the only available packages may, and usually will, mix the *existing* options so any choice must go against important preferences in some instances. Further, any prospective control is dependent on the ability of the representatives to have sufficient impact on the distributed process of decision so the directions to which they commit themselves will actually be followed. Optimism about this mode of citizen control also depends on assuming voters are able to identify their discontents, to know when their liberties have been infringed, and to trace the cause of the trouble. Finally, it supposes the consequences for them are not irrevocable, and that their votes against representatives whom they view as having imposed upon them will substitute new representatives who are able, and willing, to put things right.

It is easy for any of these conditions to fail, and, in consequence, it is an enormous illusion to think of existing democracies as providing the citizens with full enhancement of their freedom and complete control of their lives. Why then are people so enthusiastic about democracy? Why do they think of themselves as having some important form of control?

Any explanation of democracy's good press has to be multifaceted. Many citizens in contemporary democracies have little awareness of any failure to live up to the ideals of freedom and of equality in freedom. As these people think about their own lives and those of their friends, neighbors, and associates, they recognize constraints on their behavior, but they do not see these as infringements of liberty—they are nuisances, inconveniences, even real cramping of their ability to pursue their projects, but they are taken as facts of life. The student who is unable to take courses she needs to realize her ambitions, or whose progress is cramped by incompetent teaching, inadequate texts and equipment, and menacing surroundings, will feel the pinch, but she typically does not characterize it in terms of oppression or the limitation of liberty. Consequently, she does not express outrage that a democratic ideal is being violated or, at least, compromised.

A second facet of the explanation is that people most satisfied with their own liberties, people who think of existing arrangements as enabling them to participate in securing those liberties, are often unaware of the inequality with which freedoms are distributed. They use the formal guarantees of equality before the law, together with their own sense of proper provision of liberty in their own case, to conclude that democracy lives up to its ideals. If I see no problems with respect to my own freedom, and if I know that the Constitution treats me and the minority person in the rundown neighborhood as equal before the law, I may conclude that the minority person enjoys just the same freedoms I do. Of course, if I were to spend a bit of time walking the streets, looking for work and for accommodation, with that minority person at my side, I might conclude otherwise.

The principal factor in the explanation, however, is neither false consciousness nor failure of engagement with the less fortunate members of society but rather the fact that what is most salient to members of democratic societies is democracy's undoubted success with one difficulty of enormous historical importance. Democratization has been an effective response to the problem of identifiable oppressors. Even when societies have become extraordinarily complex, and when voter control is attenuated, the solutions to problems of this type endure. This is because the usual strategy for dealing with the identifiable oppressors is not just to overthrow them but to put in place clear constraints on the behavior of those who will administer after them. Constitutional provisions specify checks that limit our leaders, and, should those checks be ignored or repudiated, it is likely (but not inevitable) that the public would understand and take notice. So we reasonably expect democracy, as it currently exists, to provide protection against the most significant forms of oppression that have figured in political history. When existing democracies are compared with other regimes, the problem of identifiable oppressors looms very large, and, even in its current forms, democracy seems—as it is—clearly preferable.

Even though democracy, in its current forms, doesn't solve all the problems of promoting and enhancing freedom (and equality in freedom), it should be celebrated for what it does achieve. Reflection on the failures of full democracy in those countries that trumpet their democratic achievements most loudly can easily inspire a jaundiced attitude toward what has actually been accomplished. By contrast, citizens of nations that have recently lived through periods of tyranny—as, for example, with the undermining of German democracy under the Nazis—will rightly judge that solving the problem of identifiable oppression is no mean thing. Appreciation of that fact is compatible with realizing that there remains important work to be done.

Unfortunately, although people often lament that their vote makes no difference, there is a prevalent feeling that the limited amount of electoral control we enjoy is enough. That feeling rests on recognizing how identifiable oppression can be checked. Yet if we take the three-tier picture of democracy seriously, it is evident that the past processes of democratization need to be continued. Democracy should be regarded as a work in progress, because, important though it is, the problem of identifiable oppression, for which we have at least some effective partial solutions, is not the only problem that limits realization of the ideals that are fundamental to the value of democracy. Problems of *unidentifiable* oppression ought also to be taken seriously. These problems come in two forms, one in which an oppressive agent is good at hiding his responsibility from the citizens who might remove him, and one in which oppression exists, even though there is no person against whom the citizens can direct their repudiation. Both forms of the problem require public knowledge, if they are to be addressed. The second is pervasive in contemporary life, *precisely because of defects in the system of public knowledge we have*. Chapters 5–9 are concerned with identifying some of these defects and the forms of unidentified oppression to which they give rise, as well as offering some tentative suggestions for amending them. The next chapter prepares for that diagnostic work, by examining the ways in which our present system of public knowledge has emerged.

Chapter 4

THE EVOLUTION OF PUBLIC KNOWLEDGE

13. ORIGINS

Public knowledge has been around for a very long time. Often, it has been organized quite independently of any commitment to democratic values. Because democratic societies, and the ideals of democracy, emerged in reaction to earlier forms of political association, the systems of public knowledge embedded in them have often taken over ideas introduced in nondemocratic regimes, and some of those ideas continue to play a role in our own thinking about public knowledge. Further, the system of public knowledge we have inherited evolved haphazardly—nobody designed it or conceived it as a crucial factor in advancing democracy. There is thus no particular reason to suppose our system of public knowledge to be well designed for the functions to be served in democratic societies, or to believe our ways of thinking about public knowledge and its value to be free of elitist elements.

These are the principal themes to be defended and illustrated in this chapter. To begin, let us recognize the scope of what counts as public knowledge.

The forms of inquiry designated as "sciences," where the term is used inclusively to cover studies of art, literature and music, investigations of human behavior, culture and societies, as well as facets of the physical and organic world, play a central role in our contemporary system of public knowledge, that body of shared information on which people draw in pursuing their own projects. Because of their spectacular successes in fathoming some aspects of nature that initially seemed deeply perplexing, the natural sciences appear particularly impressive, worthy of celebration as paradigms of human knowledge. Other types of investigations should not, however, be slighted. Despite the tendency to pooh-pooh the "soft sciences," the rigor found in attributions of pictures and musical scores to artists and composers, in tracing patterns of influence among literary figures, in reconstructing past ways of life, in exposing the relations among languages, in revealing the structures of unfamiliar societies, and in a host of kindred achievements is in no way inferior to that achieved by physicists, chemists, biologists, and geologists. The contemporary system of public knowledge serves us well in many respects.

The chief aim of this book is to offer an account of how this broad system of public knowledge, with Science as a prominent part, functions to promote our values, how it contributes to and is constrained by the goals of democracy, and where our contemporary version of it might be amended to advance those values and goals. In developing this account, we need to understand the historical development of public knowledge, for it is easy to assume that features of it, developed in response to particular problems of the past, are essential features, not to be jettisoned. Later in this chapter, for example, we shall consider how an enterprise initially viewed as a private venture became central to public life.

Public knowledge exists whenever and wherever there are channels of communication through which information is transmitted from some organisms in a social group to others. Whether or not they can be said to have a language, bees have a rudimentary system of public knowledge, centered on the "waggle dances" through which they share information about potential sources of nectar and thus guide the foraging behavior of the group. Similarly, in primate troops, vocalizations serve to warn group members of a dangerous predator, gestural rehearsals instruct others in ways of performing a task, signals coordinate joint hunting (Cheney and Seyfarth 1990). Fundamental to all these activities is a primitive division of epistemic labor. Within a social group, particular individuals have access to states of the world to which others do not. Aspects of those states are pertinent to the satisfaction of the animals' wants and needs, and the transmission of information from those with privileged access to other members of the group promotes overall success.

When our ancestors acquired language, they became able to transform these rudimentary systems of public knowledge, producing something really worthy of that title. They could now distinguish ephemeral aspects of their environment from enduring features whose presence it was important to remember. It is valuable to know *now* that there is a potential food source currently in a particular place, or a dangerous animal presently lurking. Even more useful is the recognition of standing conditions—that particular places are always perilous, or that prey can usually be discovered in the vicinity of a water hole. So, besides the transmission of information to guide immediate action there arises the idea of a body of lore, to be remembered by all members of the group, and to be passed on to the young.

Public knowledge in the simplest sense consists in shared information. Human language allows for the richer resource of a corpus of statements, accepted by all adult members of the group, used to guide an indefinite sequence of actions, and carefully transmitted to the young. Human societies have possessed systems of public knowledge, in this more substantial sense, for tens of thousands of years. Like the problems arising in the early stages of the ethical project, and the rules introduced for solving them, the original content of the system of public knowledge was most likely biased toward information relevant to acquiring needed resources and avoiding environmental dangers. Besides *true* statements, correct descriptions of where to find good materials for tool making or fresh water, or of how to respond to a threatening animal, our forebears mixed into their body of lore *falsehoods*, statements accepted as public "knowledge," even though they lacked the necessary property of accuracy. They invoked personal agents who produced phenomena they could not understand in other terms—deities as sources of thunder or of the earth's renewal in spring—and increased compliance to their ethical codes by attributing to these supernatural beings a power to observe human actions (Boyer 2001; Westermarck 1924; Kitcher 2011a). The inclusive body of lore, the mix of truth and falsehood accepted by all and passed on to the young, constituted their public "knowledge"; we can think of public knowledge (proper) as restricted to the part of the lore that is actually correct.

The body of lore was generated through the efforts of many individuals—perhaps each member of the group played at least some role. At relatively early stages, our ancestors would have become aware of the possibility of *mis*information, and successful bands would have taken steps to guard against the danger of incorporating falsehood within their public "knowledge." Tacitly, they would have adopted norms of sincerity and competence. Given the original function of the ethical project, that of remedying altruism failures, insisting that band members only report their beliefs would have been a progressive step. Almost as important would be the inculcation of habits of careful inquiry. Because of the primacy of recording facts about the immediate environment, obtainable through the senses, the young would learn how to look and listen attentively, and only to deliver reports grounded in responsible perception.

The most elementary form of the division of epistemic labor starts from the fact that, in everyday activity, members of the group disperse, thus enjoying different opportunities for perceiving the local habitat. The group benefits from synthesizing the information acquired in the dispersed observations of its members. Public knowledge accrues from the efforts of differently situated individual knowers, where the role of a knower is governed by norms of truth and responsibility. Knowers are expected to supply *true* statements about the parts of the habitat they have visited, and they are expected to attend to their surroundings sufficiently to make the beliefs they acquire *reliable*.[1]

These dispersed group members are not merely expected to be knowers, reporting accurately on the areas through which they wander. They also count as *investigators*, part of whose task is to be alert for information of the sort the band takes to be relevant. Someone who returned full of vivid—and accurate—descriptions of the views available from particular perspectives, but who remained silent about the increased frequency of dangerous snakes or the falling level of the local water hole, would be living up to his responsibility as a knower but failing dismally as an investigator. Even in its early forms, the human system of public knowledge introduced norms for investigation, requiring its dispersed inquirers to discover and report those aspects of nature viewed as particularly pertinent.

How is pertinence—or relevance, or significance—determined? Through the joint deliberations of the group, who take particular questions to be important for them. Part of their body of lore consists in an understanding of their own ignorance. They learn how failure to have particular kinds of information can interfere with their purposes and cause trouble for them. Experienced difficulties confer a significance on families of questions, and group members are taught about the kinds of observations that will be particularly welcome. Their daily investigations are then expected to be sensitive to providing knowledge of the preferred types.

So far, I have described a relatively egalitarian division of epistemic labor, in which knowers/investigators are distinguished in terms of their trajectories through the habitat, rather than by distinctive talents or acquired competence. With the general division of labor, the division of epistemic labor is intensified. Band members assigned to the job of tracking prey learn which prints or fecal deposits reliably indicate the presence of which animals, which subtle modifications of the brush are produced by animal movement. As learners, they are expected to acquire these discriminatory skills; as knowers, they are expected to look attentively at the residues and traces; as investigators, they are expected to be alert to potential clues. Similarly, gatherers are to acquire and employ abilities to identify important plants and the likely presence of roots. Superimposed on the general observational competence, demanded of all, there are individual differences, separating various types of knowers. Yet, as the conception of a knower/investigator is further articulated, the original themes remain as sources of norms: true information is the goal, responsible knowers take enough care to ensure the beliefs they acquire are likely to be true, and proper investigators aim to relieve the group's ignorance with respect to those questions taken to be especially significant. With the manufacture of more complex artifacts, with the domestication of animals, with the building of relatively permanent settlements, with the construction of systems of irrigation, the labor is divided ever more finely. Under the general rubric of the knower/investigator, societies tacitly differentiate forms of expertise.

14. THE PUBLIC DEPOSITORY

Before the invention of writing, retention and transmission of public knowledge depended on the memories of individuals. As societies became larger, with a more fine-grained division of labor and a more complex lore, they may well have come to view the task of preserving their most important discoveries as requiring development of mnemonic techniques, perhaps elaborating the division of labor to provide a special place for people whose well-cultivated memories would equip them for the task of storage. Writing expands the limits, allowing for the collection of documents and the creation of a public depository.

For the past five thousand years, the public depository has been central to systems of public knowledge. The general structure of public knowledge has been defined by processes of *investigation*, *submission*, *certification*, and *transmission*, elaborated in different ways in different societies. These processes had counterparts within the earlier systems of small, preliterate groups, but the larger communities of our historical past, typically hierarchical with an elaborate division of labor, have allowed many alternative forms of refining them. At the very earliest stages, all members of the band were potential contributors, investigators whose dispersal through the habitat might produce valuable information. The types of knowledge sought were expressions of group values, accepted in deliberations in which all adults participated. Individual informants were held to norms of truth telling and attentive observation, and, unless there were grounds for suspecting their sincerity or care, the reports they brought back to the band—submitted—would be adopted as part of the group's lore—they would be certified. That lore would be passed on, transmitted in its entirety, to all.

By the time of literate societies, quite possibly much earlier, these processes have been transformed because of the specialized roles people play: not everyone counts as an investigator; investigators have special domains; certification is no longer automatic; transmission spreads knowledge to some but not to all. Just as many aspirations and desires available to (at least some) people in the earliest historical societies, those of Mesopotamia and Egypt, are beyond the horizons of the participants in the early stages of the ethical project, so, too, the types of information considered pertinent have expanded enormously. When many people live together in proximity, the differences in trajectories no longer correspond to individual perspectives that might generate useful new findings. Construction of works that play an important role in sustaining the community brings new epistemic tasks. Instead of leaving reports of the state of the environment to the wanderings of observers, the crucial human-made features require careful inspection (specialists are needed to make sure that irrigation channels are properly maintained). Chance observation is only rarely important for expanding public knowledge. For especially crucial matters, there are new techniques of appraising the reports submitted, even when those who deliver them are presumptively sincere and competent. Certification sometimes requires the production of evidence. Since public knowledge is now retained in written documents, only the literate, a small minority of the local population, can have direct access to it. Decisions have to be made about how to convey to the nonliterate many whatever information they need to perform their more- or less-specialized tasks. Even among those capable of reading, there are issues about which documents may be read by whom. The system might even, in principle, be constituted so that no single person could have access to the full range of public knowledge.

For any system of public knowledge with a public depository, we can pose a series of questions about the processes just distinguished. These questions will frame my efforts to provide a theory of how contemporary public knowledge might promote the values underlying democracy. The four processes generate a useful division.

Investigation. What types of investigation are viewed as worth pursuing? How are they supposed to proceed? What constraints govern them?

Submission. Which people are entitled to submit reports to the public depository? On which topics may they submit? How are they trained? What standards do we expect them to meet in their investigations?

Certification. What is required for submissions to be accepted for inclusion within the public depository? Which types of knowledge should be written down, which trusted to the memories of particular people? Under what conditions should statements already included in the depository be jettisoned?

Transmission. Which parts of the public depository are available to which people? How is the public knowledge needed by different people transmitted to them?

Notice that at the first stages of public knowledge, the questions do not arise, simply because very straightforward answers are taken for granted. The last section surveyed systems of public knowledge in which decisions about lines of investigation were made through egalitarian intraband deliberation, in which the questions identified as significant focused on aspects of the environment bearing on basic needs, in which the dispersal of band members made each of them a potential source of new knowledge, in which people were expected to observe carefully and report sincerely, in which reports were automatically accepted except when there were grounds to suspect informants of not living up to the norms of attentiveness and honesty, and in which information was transmitted to all. After the introduction of the division of labor, complications arose (trackers have special expertise in reporting on the movements of animals), but it is easy to imagine that some of the questions on my four-part list were irrelevant, even for most societies of the early neolithic. For the complex societies in which writing was invented, however, it is evident that all these issues arise.

In Mesopotamia and in ancient Egypt, life depended on techniques for the erection and maintenance of large structures and for the irrigation of land. Successful regulation of city life, of trade and agriculture conferred significance on questions that early systems of public knowledge could never have formulated. These societies valued investigations that produced methods of tallying, counting and measuring, keeping accounts and surveying land, tracking the movements of heavenly bodies and constructing calendars (as well as integrating their astronomical findings with attempts to predict the future by casting horoscopes). Their ability to solve quadratic equations (without modern algebraic notation!) and to recognize particular geometrical truths reveals a system of public knowledge emphasizing not only the directly practical but also theoretical techniques for addressing practical issues. Surveying and astronomy required them to specify processes for acquiring knowledge beyond those of unaided observation. Their codes of law, even in the fragmentary form in which they have come down to us, reveal an institution of private property that would constrain knowledge seeking (surveyors would have to position their instruments so as to avoid intruding on the land or property of others, to cite just one simple example).

The surviving documents make evident how Mesopotamian and Egyptian life required specialists with different training and expertise: people who scrutinized the measures used in trade, who kept the accounts, who inspected the fields and the canals, who supervised the transport and storage of goods. Performance of their various tasks was expected to accord with the procedures they had been taught, the skills they had acquired—techniques dependent, in their turn, on the knowledge of the mathematicians and astronomers who refined practices of measurement and devised the calendars. In instituting, supervising, and amending this division of epistemic labor, all the *submission* questions would have arisen.

Similarly with respect to issues of *certification*. Development of complex practices of surveying and astronomical measurement could hardly have gone forward without occasions on which people credited with expertise offered divergent conclusions. False reports were no longer simply the result of insincerity or inattention: even well-trained observers mismeasure, and their ways of going astray require diagnosis. Standards of evidence would have been articulated in the course of resolving disputes and of revising what had been taken as public knowledge.[2] It is, after all, highly implausible that the calculated calendars always worked!

Finally, in these large pyramidal societies, the esoteric knowledge of the literate would have been unavailable to the artisans, the field laborers, and the slaves. Decisions were needed to fix the programs of training people for the menial tasks required of them, as well as to determine what parts of the public depository were to be opened up to different segments of the literate elite.

Who made all these decisions? We have too little evidence to do more than speculate. Yet those involved in addressing the various issues were surely members of a tiny minority. Mesopotamian and Egyptian systems of public knowledge probably evolved over centuries, sometimes modified in large ways, sometimes nudged in this direction or that, as elite administrators recognized problems in carrying out the projects of the rulers. Complaints about trade probably stimulated renewed attention to the issue of standardizing measures (a common theme in ancient law codes), requiring adjustments in production, training, and inspection. Destructive floods prompted scrutiny of surveying techniques, refined instruments, and a respecification of roles. The bureaucrats in charge shaped a highly undemocratic system of public knowledge.

15. LATER VARIANTS

Early human systems of public knowledge, I have suggested, were linked to the practical needs of a small group, and they were egalitarian both in basing a conception of significant information on a process involving all the adult members, and in distributing knowledge throughout the entire band. By contrast, the systems we can discern in the first literate societies, in Mesopotamia and in Egypt, were profoundly inegalitarian, not only in reducing the size of the "public" that enjoyed the benefits of knowledge, but also in leaving the direction of inquiry to a tiny minority of rulers and administrators. For these systems of public knowledge it would not matter at all that some different mode of organization, some alternative choice of investigations might bring enormous relief to the vast majority of the population. Unless that rival approach was conducive to the goals for which the elite decision makers strove, it would be of no interest for them.

Many of the systems of public knowledge that have subsequently emerged have shared this jaunty disregard for what might concern the masses. They have, however, introduced a feature not obviously present in the examples so far considered: namely, the thought that certain types of knowledge might be *intrinsically* valuable, worth pursuing quite independently of any practical end they might help to bring about. So far as I can tell, this divorce of *pure knowledge* from the prospect of successful interventions in the world was not achieved in ancient Mesopotamia or ancient Egypt. In those systems of public knowledge, interest in questions accrues from the contributions answers would make either directly to practical ventures or to the construction of techniques for carrying out those ventures. Geometry grows (as its name suggests) out of the practice of measuring land; the ability to solve quadratic equations is motivated by the need to solve problems of dividing legacies.

Greek thought changes that style of evaluation. If we could fully reconstruct the system of Greek public knowledge—in, say, the Athens of Socrates and the young Plato—it would probably correspond in many respects to the public knowledge present in Mesopotamia and Egypt in the previous millennia. There would be similar divisions of specialties arising out of practical tasks: measuring land, constructing calendars, maintaining bodily health. Ideas about investigation, submission, certification, and transmission would almost certainly reveal the same patterns attributed in the previous section, with similar indifference to the state of the majority of the population. Those ideas would evolve in similar ways, as leading citizens responded to problems they recognized in their existing arrangements.

Yet there are two important features introduced in texts that have come down to us. One, manifest in Euclid's reconstruction of the scatter of geometrical results achieved by his predecessors, is the separation of mathematics from its application to measuring physical objects and the construction of a system in which claims are subjected to a demanding evidential requirement. Euclid famously gave us the axiomatic method, and the associated notion of proof, effectively changing the rules for certifying geometrical claims. The ultimate test no longer has anything to do with what happens if people make particular maneuvers around parcels of land but whether the geometer submitting the claim under scrutiny can provide a line of reasoning leading from the statements specified as axioms to the proposed conclusion. It is tempting to think that Euclid's celebration of reasoning as the standard of certification is what prompted Plato to inscribe the famous line "Let no man ignorant of geometry enter here" over the doorway of his Academy.

Perhaps there are earlier sources, but it is clear that, in the works of Plato and Aristotle, deliberate attention is given to processes of certification. Whatever procedures or rules of thumb had guided previous attempts to resolve differences among experts who made incompatible submissions or proposals to revise the system of public knowledge, there are now new branches of knowledge—a novel extension of the division of epistemic labor—charged with identifying the conditions for introducing statements into the public depository and maintaining them within it. (As with less explicit forms of attention to certification, this development is probably intertwined with issues about the resolution of private disputes, particularly in the procedures of law.) The system of public knowledge has begun to reflect on itself.

The second major change also involves an increase in self-consciousness, this time with respect to the notion of proper investigation. Socrates, Plato, and Aristotle formulated the question of how to live as the central issue of philosophy. Although they might have regarded the question democratically, as arising for all, their ways of framing it and seeking answers to it presupposed that the envisaged subject belonged to a tiny segment of the species: the well-born man of the Greek *polis*. For this subject, Plato and Aristotle offered their insightful, if schematic, answers. The good life involves activity, contributions to the community, attainment of virtue, and friendship. Beyond that lies the further achievement of understanding—possibly the capstone of all the rest.

Whether or not these visions caused large changes in the evolution of public knowledge, they have been enormously influential on subsequent *thought* about public knowledge, for they reinforce the Euclidean conception of types of knowledge that are divorced from practical problem solving, worth having for their own sake. Indeed, coupled to the emphasis on the life of the patrician male, these visions suggest parallel hierarchies in forms of knowledge and worth of lives. Most people (including most or all women[3]) cannot enjoy lives of the highest quality, and the types of knowledge they achieve are properly restricted to their mundane activities. A few have the ability to ascend to the heights of theoretical contemplation. Living on these heights confers extraordinary value: although it is important for these superior people to direct the affairs of the community, they will only reluctantly return to practical activity: the guardians would prefer to remain among the Forms.

My aim in dealing cursorily with some familiar developments in Greek thought is to recognize the important ways in which the Greeks advanced and distorted our perspectives on public knowledge. The advances come in the posing of central questions, in asking what conditions should be met for certification, and most crucially what kinds of knowledge should be sought.

Further, in placing the question of how to live at the center—not only of philosophy but of inquiry—Socrates, Plato, and Aristotle offer a framework within which discussions of public knowledge can be set. The simple systems of public knowledge of prehistory tacitly adopted a basic view of the good life as one in which elementary needs were met and were developed to promote good lives for all the members of a small band. Situated much later in the ethical project, Greek thinkers could advance a far richer conception of a worthwhile life, but they repudiated the democracy of the prehistorical approaches, directing public knowledge to the aspirations of a privileged few and aligning divisions among grades of knowledge with a hierarchy of human worth they took for granted. As §7 already suggested, we can approach the question of valuable lives (and issues of values generally) in a more egalitarian spirit, absorbing the insight that public knowledge should improve the quality of human lives while combining it with democratic values. As we attempt to do so, it will be important to guard against importing influential ideas deriving from the inegalitarian assumptions that pervade Greek approaches to knowledge.

Turn now to a second important modification of the system of public knowledge, brought about by the rise of Christianity and its dominance in western European intellectual life.[4] Like the theoretical conception embraced by the Greeks, according to which public knowledge was ultimately directed toward the attainment of the valuable life, Christian perspectives embodied a hierarchy of forms of knowledge, one that downplayed the importance of information pertinent to practical problem solving. Indeed, we can view the Christian conceptions as endorsing the centrality of the Greek question—"How to live?"—while suggesting a radically different answer. The valuable life is one directed toward God, and its ultimate consummation, as Augustine makes clear in the powerful crescendo that ends *City of God*, lies in union with God in the hereafter. Interestingly, although it retains the Greek notion that some types of knowledge are suitable only for the few, the Christian conception restores some of the democratic themes the Greeks had abandoned. Knowledge of the highest type is knowledge of God, and although parts of this knowledge—detailed understanding of the mysteries of the Trinity, say—cannot be appreciated by the masses, the most central and significant truths can be known by all (except the feeble-minded). Even the humble plowman, and his equally humble wife, can learn that their world was created by God, that they have been born into sin, and that they are redeemed by the sacrifice of Jesus. Although they cannot fathom the details of the relations between the persons of the Trinity, St Patrick can show them the shamrock.

Public knowledge (at least officially) is centered on the most crucial truths of religion, statements that characterize God as creator and sustainer, that describe the predicament of human beings as fallen creatures, and that tell the story of our redemption. Further explorations in theology should be undertaken to deepen the understanding of these fundamental tenets—perhaps because it is good for the elite to enjoy the theoretical contemplation of these deep facts about the creation (a reworking of the Platonic-Aristotelian celebration of "pure knowledge" within the specifically Christian context), perhaps because the profounder comprehension will enable those who enjoy it to convey the central message more clearly and more vividly to the masses. Other forms of inquiry, including those that build on knowledge acquired in the ancient, pre-Christian world (for the most part transmitted through Islamic intermediaries, who had often made important extensions to it), obtain their significance through the contributions they make to the understanding of the deity and his virtues. Questions about that significance are much debated: Does pagan knowledge have any value? Is it important to extend ancient treatments of natural phenomena (studies of local motion, for example)? Is mathematics—beyond arithmetic and the system of Euclidean geometry—more than a curiosity, a set of diverting puzzles?

Christian conceptions of public knowledge are embodied in the activities of monasteries, schools, and universities—most evident in the divisions of the curriculum. The official hierarchy of types of knowledge is partially reflected in social divisions, in the relative status of people who figure differently in the division of epistemic labor. Reflection is *only* partial because of the pervasive role of the church in everyday life. Practical administration requires knowledge of the law, and success is dependent on skills in logic and rhetoric. Outbreaks of disease serve as reminders of the importance of medical investigations. Thus, independently of their contributions to the understanding of God, some divisions of public knowledge gain elevated status. Others, like mathematics and the study of motion, seem to have little impact on comprehension of the deity or on practical affairs. Only when it becomes plain that reform of the calendar is urgently required would it seem appropriate to request a Polish monk, trained in canon law, to turn his attention to these lesser parts of the public knowledge system.

The hierarchy of types of knowledge leaves an imprint on ideas about certification, one that endures to this day. On the one hand are investigations of major religious questions, pursued by the most elevated intellects with diligent attention to a long tradition of texts and authorities; on the other, the relatively low-status inquiries into natural phenomena, with their less important—but recalcitrant—questions and their inconclusive answers. Where claims made in these different areas conflict, it is very natural to suppose that the work of textual interpretation is more rigorous and sure-footed, that the methods it articulates and uses are more likely to deliver truth, and so to view the appeal to religious authority as having the power to override whatever evidence is drawn from a mix of observation and mathematical reasoning. Even when the successes of the new sciences began to suggest that the evidential force of inquiries in these areas had been underrated, it remained difficult to dismiss the practice of interpreting sacred texts. Surely the deliverances of religious tradition constitute part of the *total* evidence?

The Greek and Christian conceptions I have sketched reveal alternative, internally coherent ways of thinking about public knowledge. I now want to explore how—partly by reaction, partly by incorporation—our institutions of Science emerged from them.

16. FROM PRIVATE TO PUBLIC

The most important feature of Science is that it was not planned.[5] To be sure, many of the early pioneers who transformed what we now call the "sciences" (which they typically thought of as "natural philosophy") had strong opinions about how individual or collective inquiry should be carried out. Descartes, Galileo, and Newton all gave serious thought to questions of method (conceived as exemplified by the knowledge-seeking activities of an individual), and Bacon wrote at length about what was required of an individual investigator and about the proper organization of a community of inquirers. Lord Chancellor that he was, Bacon might have harbored ambitions of displacing the public system of knowledge, building something sounder for directing the activities of a nation. Others, even those inspired by Bacon's own vision of individual and collective inquiry, simply stepped outside the existing public knowledge system, with its institutions and constraints. They wanted to explore different things and to do so in different ways, to pose their own questions and satisfy their own curiosity. Insofar as their venture was a collective one, it would be undertaken for the benefit of a small group of comrades.

In 1660, shortly after the restoration of the British monarchy, a group of brilliant gentleman scholars persuaded the new king to lend his name to their endeavors. Thus, the Royal Society was born, and, so the story goes, it quickly gave amusement to the royal patron. Learning of Robert Boyle's experiments with the air pump, Charles II is said to have laughed at the idea that the learned gentlemen were "weighing the air." Little more than a decade later, in 1676, the antics of the Royal Society were sufficiently familiar to the public that Thomas Shadwell could offer the London stage a successful comedy, *The Virtuoso*, whose central character, Sir Nicholas Gimcrack, introduced a new word into the language. The following exchange is representative:

> SIR NICHOLAS: I content myself with the speculative part of swimming; I care not for the practice. I seldom bring anything to use: 'tis not my way. Knowledge is my ultimate end.
>
> BRUCE: You have reason, sir. Knowledge is like virtue, its own reward.
>
> SIR FORMAL: To study for use is base and mercenary, below the serene and quiet temper of a sedate philosopher.
>
> SIR NICHOLAS: You have hit it right, sir, I never studied anything for use but physic, which I administer to poor people.

Eleven years before the publication of Newton's *Principia*, the members of the century's most distinguished collective for scientific inquiry were not perceived as having a social role. They were members of a club—"gentlemen, free and unconfin'd" as they described themselves—and those suspected of "base" motives, men who might raise and pursue questions in the interests of "trade," were to be excluded, as were cranky unclubbable men (Hobbes) and all women. (After an amendment to the Royal Society's statutes was passed in 1944, the first female fellows were elected in March 1945.)

No social role, no social responsibility. Not only were many of those who participated in the early Royal Society outside the official institutions of the public knowledge system—the universities—they were often contemptuous of those institutions. Even the few, like Newton, who held professorships would not have seen the activities of the new group of inquirers as competing with the extant public knowledge system. They saw themselves as *autonomous*, ungoverned by the norms for inquiry and the dissemination of information prevailing within universities and kindred venues. "Free and unconfin'd," they could pose whatever questions they wanted and make up the club rules as they—collectively—thought best.

Science, in the singular, the social institution we now have, descends from a series of originally *private* projects that have gradually come to seem useful, valuable, the high point of human achievement, even indispensable to modern societies. Science has become embedded in our public knowledge system, not only an important segment of it but, through the prestige acquired by its most spectacular successes, a part that sets standards to which other forms of inquiry should aspire. (As noted in §13, many other areas of investigation are often seen as poor relations, despite their solid achievements.) The transition spans the centuries. Only in the nineteenth century did scientific research obtain a central place in the universities; only in the twentieth did the investment in scientific inquiry become perceived as important for national strength, and education in the sciences come to appear essential.

Science has evolved by happenstance, its position within various societies shaped by contingent events and the opportunities they offered. Its structures and norms have come about partly through the retention of ideas present at the beginning, considerations that struck the private gentlemen as appropriate, partly through the growth of allied institutions (the development of the German research university and its transplantation to the United States), partly through urgent demands (particularly at times of war). Thought about Science, as about public knowledge generally, retains elements from the Greek and Christian conceptions of higher and lower forms of knowledge—evident in Sir Nicholas Gimcrack's attitude toward "the speculative part of swimming." Where the systems of public knowledge present in the bands of the Paleolithic have an evident structure and coherence, where the systems of ancient Greece and of Christianized Western Europe have a more or less explicit rationale, the elements of our own system, with institutionalized Science as a prominent part of it, have emerged contingently and haphazardly. Not much time, if any, has been devoted to wondering about how public knowledge might be shaped so as to be good for democracy. We lack any convincing theoretical conception of how Science contributes to valuable goals.

This last judgment may seem overstated. For, in the past century and a half, many prominent thinkers have offered suggestions about the roles of

Science and of scientists. Francis Galton (1875), Darwin's cousin and the apostle of eugenics, offered the image of the community of scientists as a secular priesthood, whose responsibility is to guide human intellectual and moral life. His conception remains influential, surfacing in the rhetoric of such recent and contemporary champions of science as Carl Sagan, Richard Dawkins (1998, 6), and E. O. Wilson, who write of our progress in "disenchanting nature" and ask if it is not a "noble, an enlightened way of spending our brief time in the sun, to work at understanding the universe." Remarks of this sort echo the Aristotelian conception of theoretical contemplation as an important way of fulfilling our humanity, although the writers I have cited divorce the ideal from Aristotle's elitism (which the gentlemen of the seventeenth century would probably have shared), passionately dedicating themselves to making the joys of scientific understanding widely available.

On the other hand, a prime engine in the institutionalization of Science required a very different image. In the wake of the Second World War, Vannevar Bush brilliantly developed a utilitarian case for public support of Science. Scientific research is not, as Charles II and Thomas Shadwell apparently thought, a useless exercise for dilettantes, nor is it simply a matter of pursuing truth for the sake of disinterested understanding (achieving the wonderful ends celebrated by Plato and Aristotle). Inquiring into nature, even when it seems remote from practical matters, is a superb strategy for future advances that will transform ways of meeting basic human needs. Down the road it will offer an increased food supply, treatments for diseases, greater mobility and more secure shelter, and improved means of defending ourselves from enemies. Investing in research, especially "basic research," is to harvest seed corn from which we may hope to feed well in the future. Interestingly, whereas Bush's brief for publicly supported Science reverses the view of the seventeenth-century savants who resisted directing their inquiries toward realizing profit ("trade" was viewed with anathema), it preserves the idea of scientific autonomy: the public is to provide, but the community of scientists is to decide how to assign the resources delivered among potential investigations.

Optimistic visions like those just reviewed contrast with others that view any system of public knowledge as potentially oppressive. In some circles, Science comes to be seen as the heir of the Church or the Party, the beneficiary of an ideology that constrains people's thought and their lives. Opposing its regimentation, Michel Foucault takes as his task an "insurrection of knowledges against the institutions and against effects of the knowledge and power that invests scientific discourse" (1980, 87). Even if you resist Foucault's rhetoric, you might appreciate a real insight here: does Science (or public knowledge) play the needed role of disclosing forms of oppression that would otherwise go unidentified (§12)?

We have inherited an institution with many disparate elements, pointing toward rather different ends. Intelligent people, reflecting on Science, find this or that aspect salient, using their observations to sketch an account of the values and functions of scientific inquiry and knowledge. It is not simply that the pictures they offer are very different, but there seems to be no framework for a continuing conversation. The popular allergy to probing value-judgments, to making values explicit and trying to understand and defend them, generates a collection of rival visions with little chance of synthesis. Consequently, although previous systems of public knowledge allowed for coherent theoretical treatment—and, for all the brevity of my presentations, the last section offered outline accounts of two historically important conceptions—we lack anything similar for contemporary public knowledge and for Science as a prominent part of it. The upshot, as chapter 1 argued, is a sequence of challenges to particular lines of scientific investigation and an erosion of the authority of Science.

It is now possible for me to draw the threads of my previous discussions together and to be more precise about the kind of theory I envisage. The central task is to understand what valuable ends a system of public knowledge, with Science at its center, might promote, how our existing system succeeds with respect to those ends, and how its functioning might be further improved. To undertake that task requires an explicit commitment to discussing values: chapter 2 explains the terms in which I plan to proceed. Additionally, it is important to understand the kind of society to which our institutions are to be adapted and specifically its self-conception: because the contemporary societies in which Science is embedded think of themselves as democracies, it was necessary, in chapter 3, to scrutinize the requirements of democracy (and, as I argued, to relate those to the approach to values delineated in chapter 2). The discussions of the present chapter have looked quickly at a number of very different ways in which public knowledge might be organized, not only because some elements of past perspectives probably linger in our present practices, but also because consideration of those rival approaches exposes questions it will be important to address. I shall close this chapter by developing this last point, thereby setting an agenda for its immediate successors.

Any institution of public knowledge has to do many things. It must retain the contributions of the past, organizing those contributions in ways that make it possible for whatever class of people it is to serve to gain access to whatever types of information are deemed suitable for them. It must identify the important questions to be pursued at the next stage of inquiry. In doing so, it must consider ways in which those questions might be pursued—for in the absence of such consideration it will be hard to determine which issues count as the most significant. It must decide which groups of individuals are to be entitled to submit answers to significant questions, and when the answers are sufficiently well supported to be certified—placed in the public depository, inscribed "on the books." It must decide how much discussion of established knowledge is possible. It must also resolve how much heterogeneity of opinion is valuable, and how much diversity among the community of inquirers it is good to encourage. Finally, it must decide how to deploy the knowledge available, and particularly what is to be done when urgent problems arise before fully certified solutions are available.

Other issues are entangled with those I have listed. Who belongs to the public to be served? How much dissent within the public is valuable? When should criticism come to an end? To what extent should knowledge be allowed to be private? In what ways is the pursuit of knowledge to be supported?

The four processes distinguished in §14 provide useful divisions for approaching these various questions. The next chapter will focus on *investigation* and *submission*, considering how inquiries might be assessed for significance and how they might be pursued. Chapter 6 will look at *certification* and at the conditions under which certified knowledge can guide democratic policy making. Chapter 7 will consider *transmission*, both in terms of the application of certified knowledge and the transmission of knowledge to the public. Chapter 8 will explore issues about diversity and dissent. In light of these discussions, it will be possible to return to the issues raised in chapter 1, and to appraise some of the debates about Science that currently occupy us.

As we have seen in this chapter, questions of the types just listed can be answered in many different ways. Public knowledge does not have to embody any democratic ideal: it can resolve the tension between democracy and expertise in favor of expertise (as Plato did). It does not have to suppose there to be any intrinsic value in pure understanding, independent of some contribution to practical problem solving. It can identify the "public" as a privileged group of people, as the citizens of a particular state, or as the entire human population. The system we have inherited may turn out to accord with democratic ideals—but, since it has not been planned to promote those ideals, that would be a happy accident. This book is written in the conviction that it is worth taking a look.

Chapter 5

WELL-ORDERED SCIENCE

17. SCIENTIFIC SIGNIFICANCE

Public systems of knowledge expect investigators to contribute new statements to the public depository. They expect the contributions to be worth having. It is evident that many things we might come to know—*easily* come to know—would be poor targets for investigation. My speculations about Paleolithic systems of public knowledge supposed that detailed reports of the vistas from places in the surrounding environment would not be welcome news. Similar points apply today. There are enormous numbers of ways of describing the parts of the world we visit and vast numbers of true descriptions that could be supplied—about temperature, color, spatial relations, the number of objects of specific types—which only the monomaniacal would find interesting. Nor is truth always our concern. Sometimes an approximation, a statement recognized as false but "true enough" will serve our purposes (Elgin 2004).[1]

If contemporary Science, and the public system of knowledge in which it is embedded, is to serve the purposes of citizens of a democratic society, what kinds of investigation should be pursued? Although it is easy to conceive the public depository as a collection of statements, many kinds of investigation aim at nonlinguistic products: researchers seek new molecules, new organisms, new drugs, new instruments, new techniques. Think of a *problem for investigation* as arising when some entity of a specified type is sought. Problems worth pursuing can be labeled as *significant*. Those problems are *adequately solved* when an item is produced that is close enough to the type sought to serve the purposes that confer significance on the problem. If the problem is to answer a question, an adequate solution is a statement "true enough" to enable those who have it to achieve whatever ends made the question significant. If the problem is to produce a new vaccine, an adequate solution is one providing acceptable protection against the pertinent disease. If the problem is to develop a new technique, an adequate solution is one allowing people to proceed sufficiently successfully in the contexts of intended use.

The first task of this chapter is to explain a notion of scientific significance that will fit with democratic values. It should be evident that the notion of significance is value-laden, and the explanation I shall develop will flow directly from the approach to values offered in chapter 2. Scientific significance accrues to those problems that would be singled out under a condition of *well-ordered science*: science is well ordered when its specification of the problems to be pursued would be endorsed by an ideal conversation, embodying all human points of view, under conditions of mutual engagement.[2] This understanding of scientific significance will require development and defense. Before proceeding to the explanations, however, it is worth being fully explicit about why popular treatments of significance, usually under the rubric of the goals (or ends) of Science, are deficient.

Scientists and philosophers often declare that the aim of the sciences is to provide us with a complete true story of our world. Plainly that cannot be right. There is some large infinity of languages people might adopt for talking about nature: think of the myriad ways in which the boundaries of objects can be drawn and in which objects can be grouped together. For each of these languages, there is a large infinity of true statements about the cosmos. Given these elementary facts, it is not obvious that the notion of the "whole truth" is coherent, and, even if it is, it is surely beyond human formulation or comprehension. Moreover, well-established parts of physics inform us that some parts of the universe are completely inaccessible to us: regions outside our light cone are a prime example. These losses are not serious, for virtually all of the "whole truth" lacks any interest for anybody (think, for example, of the large infinity of truths about the areas of triangles whose vertices are three arbitrarily chosen objects). Supposing that Science aims at the complete true story of the world is as misguided as the suggestion that geography seeks to draw a universal map, one revealing every feature of the globe.[3]

Behind the casual proposal that Science aims at the "whole truth" is a more plausible idea. Thinking of the sciences as primarily in the business of providing theoretical understanding (a legacy of ideas about the value of knowledge that have been influential since Plato and Aristotle, ideas that clearly moved the "gentlemen" who formed the scientific societies of the seventeenth century), scholars envisage an objective agenda, set for us by nature, to which human beings, as cognitive agents, should respond.[4] Unfortunately, nature's agenda setting is a metaphor, one that dissolves under scrutiny. The only plausible way to give it substance is to substitute a different metaphor, suggesting that Science seeks a full inventory of the "laws of nature."

Scientific inquiry does sometimes look for generalizations—and with good reason. Knowing something general can be valuable, for you may be able to use the generalization to answer many significant specific questions: Newton's second law can be applied to different dynamical systems, the generalizations embodied in the genetic code can be used to make predictions about lots of amino acid sequences, and so forth. Yet this banal point leaves plenty of room for variant ideas about the aims of the sciences. Are generalizations the *only* significant statements? Surely not. We sometimes take questions about specific things—earthquake zones, particular disease vectors—to be prime targets for scientific research. What kinds of generalizations count as *laws*? This is a perennially difficult philosophical question, and none of the (variously problematic) attempts to answer it explains why the laws of nature might be specially worth knowing.

Here again, thinking about Science and its goals is tainted by the residues of conceptions people have long discarded. The predilection for talking about "laws" of nature (which sits oddly with the haphazard ways in which particular scientific contributions are labeled as "laws," "rules," "principles," and "theories") was entirely explicable at a time when investigators thought in explicitly theological terms, seeing deep generalizations about natural phenomena as expressing the decrees of a law *giver*. Copernicus, Kepler, Descartes, Boyle, and Newton imagined their research would reconstruct part of the divine rulebook used by the Creator in setting up the show and that the reconstruction would enable people to "think God's thoughts after him" (Burtt 1932). *They* could have answered any challenge to explain why finding the laws is a worthwhile goal, but, when the theology drops away, we are left with the idea that our universe operates *as if* it has certain fundamental rules. Why should those rules be paradigms of significance?

The best hope for identifying goals for Science that will provide an "objective agenda" and thus steer clear of worrisome value-judgments is to try to build on the banal point that generalization is typically useful. Suppose Euclid's brilliant strategy for geometry works for Science as a whole. There is some manageable collection of fundamental laws, from which all generalizations about nature flow: the fundamental laws are the first principles of a "theory of everything." Whatever particular explanations or predictions are needed can be generated from this theory by conjoining the principles of the grand axiomatic system to statements about specific conditions. Science supplies an all-purpose instrument, available to anyone to understand or foresee whatever things interest him. Judgments about what matters, what is significant, can thus be left to the idiosyncratic interests of the users of a universally applicable tool.

Although this picture of the unity of Science has attracted many adherents, it cannot be sustained. For the presupposition of a series of reductions, available "in principle," breaks down once the likeliest candidates are examined closely. Classical genetics and molecular biology are well-developed sciences, and it is plain that the latter has been immensely valuable in refining views about hereditary phenomena. Despite the insights provided by chemical understanding of biologically important molecules, it is false to suppose that *every* significant generalization can be derived and explained within molecular biology. Consider, for example, the principle that genes on different chromosomes assort independently at meiosis. This cannot be derived from principles of molecular biology, since there is no way of singling out, within the language of molecular biology, all and only those entities that count as genes (or as chromosomes—or what processes count as meiotic divisions). Further, even if this obstacle were overcome, the *explanation* of independent assortment focuses on the general structure of the process of meiotic division: at meiosis, homologous chromosomes are paired, and, after exchange of genetic material between homologues, one member of each pair is passed on to a gamete. Because of the way the pairing and separation works, genes on different chromosomes are transmitted independently (Kitcher 1984; 1999).

The *single* axiomatic system, adequate for everything, is unavailable. Perhaps Science could settle for less—for a bundle of sciences, each coming with its fundamental laws. Less tidy than the imagined (and imaginary) theory of everything, a manageable bundle would still provide a universal instrument and would thus expel value-judgments from Science and leave them to the variant tastes of individuals. Even among the natural sciences, there is little hope this retreat will succeed: in many areas of biology and the earth and atmospheric sciences basic laws are hard to come by. If the realm of Science is extended into psychology, economics, and sociology, the prospects are dimmer still. Moreover, there is no basis for thinking that as new areas of inquiry are developed, they will yield further small clusters of generalizations. On the basis of the sciences so far achieved, different areas of inquiry are likely to be disparate, some allowing a small set of powerful generalizations, others requiring ramified ways of treating a large range of variation among cases (Cartwright 1999).

The resolute efforts to ban value-judgments in considering the ends of the sciences have obscured the fact that, like earlier societies that have constructed systems of public knowledge, we expect inquiry to help us with *particular types of problems*. As with our predecessors, some of these problems are practical: in those domains of inquiry that bear on medicine, agriculture, and responding to the challenges of the environment, a search for some cognitive benefit—detached understanding, say—is not primary. The aim is to grow healthy crops under adverse conditions, to find ways of curing or treating a serious disease, to know in advance the path of the hurricane or the site of the earthquake. Generalizations, laws of nature are welcome to the extent we can discover them, and all the better if they enable us to deal with a wide spectrum of cases—but if we can reach our practical ends without them, that is good enough.

On the face of it, there are also sciences that do seek understanding for its own sake. Even if an understanding of the origins of our species (or of life, or of the solar system) offered no practical payoff whatsoever, many people would still view it as valuable for its own sake. How the various hominid species evolved, and how *Homo sapiens* came to be the last one standing, is something they want to know, something about which they are curious. We should no more ignore the fact that some great scientific achievements answer human curiosity than we should slight the impact scientific knowledge has on human lives. There are three simple, but misguided, suggestions about the aims of Science and thus about the proper pursuit of scientific research.

A. The aim of Science is to discover those fundamental principles that would enable us to understand nature.
B. The aim of Science is to solve practical problems.
C. The aim of Science is to solve practical problems, but, since history shows that the achievement of understanding is a means to this end, seeking fundamental principles (generalizations, laws) is an appropriate derivative goal.

There once was a system of public knowledge in which A made excellent sense, namely, the Christian conception reviewed briefly in the last chapter (§15). Devout investigators who accepted knowledge of God, of his attributes and purposes, could think of their own inquiries as disclosing the ideas realized in the Creation. To understand nature, to reveal its most fundamental modes of organization, was to decipher a second text bestowed upon us by a wise and benevolent deity. Besides the revelation of the scriptures, humanity could study the Book of Nature, enhancing our appreciation of the glory of God. The pious Robert Boyle endowed the lecture series named for him with precisely that end in view.

As I have argued, without the theological backdrop—and the associated conception of public knowledge—A has no plausibility. Even the slightest sympathy with pragmatism (in either the philosophical or the everyday sense) will recognize circumstances in which the esoteric interests of scientific specialists ought to give way to the urgent needs of people who live in poverty and squalor. By suggesting that the areas in which pure understanding is sought for its own sake stem from widely shared forms of human curiosity, I have not vulgarly thrown pure theory into the balance with applied sciences: *it was already there*. Systems of public knowledge, including the ones that treat Science as a central part, cannot avoid value-judgments about what is significant and what is not, and some of those judgments turn on weighing the competing claims of pure understanding and practical problem solving.

Precisely because I see a competition here, I cannot opt for the simplistic pragmatism that would repudiate A in favor of B or C. Focusing just on the practical (as B recommends) would often be misguided, inefficient, or unproductive, as champions of C will point out. Often the best route to potential gains down the road is to investigate quite recondite questions: Thomas Hunt Morgan's wise decision to postpone consideration of human medical genetics and concentrate on fruit flies prepared the way for the (ongoing) revolution in which molecular understandings are transforming medical practice. Nevertheless, C inherits a major error from B by failing to recognize the ways the ethical project has expanded the scope of human desires, equipping us with richer notions of what it is to live well, ideals that include, even if they should not be limited by, the attainment of understanding and the satisfaction of curiosity for its own sake.

Not only must our system of public knowledge make value-judgments but the questions to be confronted are *hard.* They involve weighing two types of goods it is very difficult to reduce to a common measure: value accrues to answering a large question that arouses our curiosity; it is also valuable to advance human welfare. There is no escape from the balancing business, and it is not easy to see how to begin the balancing. That, of course, motivates the attempted solutions I have been criticizing, the efforts at uncovering an "objective," "neutral" agenda for Science, in which an all-purpose instrument is devised and given to individual people to use as they think fit. If that could be done, we could avoid the challenge of weighing variant values and idiosyncratic tastes.

Once we see that the challenge has to be faced, and once we recognize its form, we should appreciate the many ways in which balancing is required. If you take seriously the idea that Science is for the *human* good (not American good, not the good of intellectuals, not the good of affluent, well-educated people), you will see how it is necessary to balance the interests of very disparate groups of people. There will be issues about the schedules on which problems are to be tackled, about whether strategies offering long-term success are to be preferred or whether some issues are so urgent we cannot wait. Judgments of significance involve a multidimensional balancing act.

How, then, should we do it?

18. WELL-ORDERED SCIENCE: EXPLANATION

Many theorists who have reflected on values, or on the values realized in gaining knowledge, would answer the question directly. Some would embrace one of the positions rejected in the previous section, announcing the overriding importance of knowledge of the deity, or of theoretical contemplation, or of increasing the sum of pleasure-minus-pain across the class of sentient beings. Others would strive to give an authoritative answer, based on a scheme for placing different types of consequences on a single scale and assessing their relative weights. *Any such direct resolution is at odds with the approach to values outlined in chapter 2, an approach to be applied to the issue at hand.* According to that approach, the answer is not for any single person—not even an insightful religious teacher or a clever philosopher—to determine. Individuals can make *proposals*, but the only authority in this arena derives from a conversation. Tentative proposals about the character of the conversation are valuable to the extent they facilitate discussion. That is the way of the ethical project.[5]

Section 7 proposed that we renew the ethical project by eliminating particular accretions it has taken on during its tens-of-thousands-year history, emulating the focus of the early stages, but scaling up to recognize that the "band" in which we live is the human species. The suggested overarching conception of the general good is a state in which all people are offered serious, and equal, opportunities for worthwhile lives, where worthwhile lives are understood in terms of free choice of projects, some of which involve interactions with others. Decisions about norms and values should accord with those that would be reached in a panhuman conversation under conditions of mutual engagement.

Chapter 3 suggested that these proposals elaborate a deep democratic ideal, one that views democracy as important because of its promotion of varieties of freedom, distributed equally across humankind. One obvious way to approach some of the issues of balancing that underlie judgments of significance is to appeal to democratic principles. You start from the picture of a population with different aspirations and interests, and suppose scientific significance involves integrating these diverse elements, producing some kind of collective good. If you are tempted by the thin view of democracy (which chapter 3 attempted to transcend), seeing democracy as residing in the possibility of free elections and voting, you will suppose the apt standard for scientific significance is majority vote: each member of the population thinks about the investigations she would like to see go forward, and everybody then casts a vote.[6] Many people, especially scientists, worry that this would be an extremely bad procedure for arriving at judgments of significance. They point out, quite reasonably, that ascriptions of significance achieved in this way would favor short-term practical inquiries over research of long-term significance, that the emergent research agenda would be myopic and probably unfruitful. From the first discussions of the public role in decisions about what kinds of science should be done, scientists have taken steps to avoid confinement by public control. Vannevar Bush's masterstroke was to argue for a framework of decision making that ensured the reins could never be pulled tight—the attributions of significance were to be the province of the experts.

From the perspectives of chapters 2 and 3, both polar positions—the appeal to the thin ("voting") conception of democracy and the expert reaction to it—are completely misguided. Built in to the ideal of discussion under mutual engagement are cognitive and affective constraints: instead of myopic voters choosing in ignorance of the possibilities, and of the consequences for others, completely absorbed in their own self-directed wishes, the ideal conversationalists are to have a wide understanding of the various lines of research, what they might accomplish, how various findings would affect others, how those others adjust their starting preferences, and the conversationalists are dedicated to promoting the wishes other participants eventually form. As we shall see shortly, there is no reason to suppose that judgments of significance achieved in this way should cause scientific shudders.

The trouble with putting judgments of significance to majority vote is not the *democracy* but the *vulgarity* of the view of democracy it embodies. The reaction—to place decisions about significance in the hands of experts—might well be superior to the tyranny of ignorance that vulgar democracy would likely produce, but it arrogates to the expert community a judgment about values it is unqualified to make. It is another distortion of the ethical project, another mode of undermining the authority of a conversation among affected parties and replacing it with the *illegitimate* authority of a group. Anyone tempted to acquiesce in that authority should seriously consider the virtues of Plato's *kallipolis*, in which decisions are similarly left to the judgment of the wise.

These strong charges rest on recognizing how the conception of an ideal discussion under mutual engagement offers a better standard for scientific significance. Familiar features of everyday decision making provide motivation. Most adult members of large societies face the general problem of balancing one sort of activity against another, apportioning time to a variety of worthwhile projects. Although we sometimes give weight to one way of spending time on the grounds that it will enhance other enterprises in which we take an interest, there are many instances in which we identify two very different sources of value and are unwilling to slight either completely. In reflecting on our apportioning decisions, we recognize that we sometimes make them badly. Over the course of our lives, we develop strategies for avoiding the kinds of mistakes to which we are most susceptible.

Whatever skills we develop are put to work in joint decision making with family and friends. We would think it absurd to make plans by immediately drawing up a list of options, taking a vote, and proceeding in whatever way achieved the majority. Better to talk first. An outcome that represents the collective will should be based on genuine appreciation of the possibilities, on recognition of the felt needs of others, on understanding how the options would bear on those needs, on tracking the ways in which all of us modify our views in learning about what others want, and on a determination to avoid an outcome that someone would find unacceptable. Except when something has to be done very quickly, it is worth taking time to explore what others know and what others want. If voting ever occurs, it is as a matter of last resort, when we reluctantly agree that consensus is impossible.

These important characteristics of responsible decision making, both in balancing our own lives and in joint activities with those about whom we care, are reflected more precisely in the conditions of mutual engagement (§7), and those conditions yield my ideal of well-ordered science. A society practicing scientific inquiry is well ordered just in case it assigns priorities to lines of investigation through discussions whose conclusions are those that would be reached through deliberation under mutual engagement and which expose the grounds such deliberation would present. The society is likely to contain many different views about how the course of inquiry should now proceed; some, maybe most, of these perspectives may be sadly handicapped by ignorance of the state of the various sciences. Given the cognitive requirements on mutual engagement, that must be corrected. So we should suppose that, in an ideal deliberation, representatives of the various points of view come together and, at the first phase of the discussion, gain a clear sense of what has so far been accomplished and of what possibilities it opens up for new investigation. Those who have been addressing the technical questions of particular fields explain why they regard certain findings, particular products of research, and various currently unanswered questions to be significant. Sometimes they suggest that a question has intrinsic interest, that answering it would satisfy human curiosity; on other occasions, they relate how the answer has practical potential; on yet others, they mention both kinds of factors. At the end of this explanatory period, all the participants in the deliberation have been *tutored*; they have a picture of how the various fields of inquiry are currently constituted, in the sense of seeing how significance is taken to accrue to projects researchers have undertaken in the past and a range of options now available.

At this stage, the deliberators assess those options by voicing their own preferences. Initially, their preferences will embody their individual points of view, already amended from their previous untutored state through a clear appreciation of what might advance their personal goals. The views they set forth thus reflect their newly achieved awareness of the current state of the sciences. As each listens to the attitudes of others, preferences are further modified, since each wishes to accommodate the others, insofar as this is possible and, especially, to avoid outcomes that leave some of their fellows completely unsatisfied. Where there are difficulties and disagreements, they use the processes of mirroring, primitive and extended, to consider their potential actions from a wide range of perspectives.

As they look toward the future, their assessment of consequences, for themselves and for others, will sometimes require judgments about the likely outcomes of pursuing various investigations. Here they will need the testimony of expert witnesses. The pertinent experts are selected by following chains of deference: all participants initially defer to the community of scientists; within this community, there is deference to fields, subfields, and ultimately to individuals. Sometimes, of course, there will be serious controversy, and the chains will bifurcate. When there are rival "experts" making incompatible forecasts, the entire package is presented to the deliberators, together with the grounds on which the various estimates are made, as well as the track records of those who make them.

Conversation may end in one of three states. The best outcome is for the deliberators to reach a plan all perceive as best. Considering the conduct of inquiry within the entire spectrum of their society's projects, they judge a particular level of support for continuing research to be good, and they agree on a way of dividing the support among various lines of investigation. Second best is for each person to specify a set of plans he considers acceptable, and for the intersection of these sets to be nonempty. If there is a unique plan in the intersection, it is chosen; if more than one plan is acceptable to everyone, the choice is made through majority vote. The third option occurs when there is no plan acceptable to all and when the choice is made by majority vote. That is a last resort for expressing the collective will.

Three points should be obvious. First, the procedure outlined applies to the problem of assessing scientific significance, reflecting the general approach to value-judgments developed in §7. Second, that procedure idealizes mundane occasions of what would be viewed as good decision making. Third, any *actual* conversation of this type is impossible. This last fact may incline you to think it absurd to approach scientific significance as I have done. Understanding an *ideal*, however, can sometimes help us to improve our practice, and this is the hope of my proposal. The next section will attempt to disclose some reasons for hope.

So far, the ideal is not fully specific, since it refers, vaguely, to the range of points of view present in a society without saying how large or small this society may be. Chapter 2 favors a *broad* conception, one that would require scientific significance to be assessed by considering all the alternative perspectives present in the human population, including those of people yet unborn. Of course, those future perspectives cannot be known with any precision, but they can be estimated by further ventures in mutual engagement, by sympathetic understanding of their attitudes toward particular world-conditions we might bequeath to our descendants: it is hardly speculative to suppose they would be indifferent to a world in which violent disruptions of agriculture and water supply were commonplace. Plainly, one could draw boundaries more narrowly. One obvious way to do so is to propose that societies are identified with nation-states: the scientific practice of a particular nation is well ordered just in case its judgments about significance reflect those reached in an ideal deliberation embodying *all and only the perspectives present in that nation*. There are many others: you could confine the deliberators to some group of scientists, or to the community of tycoons, or to people who score above a particular value on some test purported to measure intelligence, or to "gentlemen, free and unconfin'd."

The possibilities just listed are intended to be unattractive. Not only is it an obvious retreat from any ideal of democracy to leave the judgments to the few, but it flies in the face of the significance of Science as an institution. Seventeenth-century gentlemen could pursue what they liked, for they had no basis for understanding how enterprises descending from theirs would transform the world, not just the world of their comfortable environs but the world inhabited by *everybody*. Future perspectives deserve representation because we know how consequential present decisions are for the people who will come after us. The choices we make will have important effects on the problems to be confronted tomorrow. Human needs arise in an environment, and the environment, since the seventeenth century, has been increasingly shaped by the particular course inquiry has taken. How could value-judgments ignore the standpoints of future people—how could ideal deliberations leave them out? By the same token, how could the scope of the conversationalists be restricted to a privileged subset of those who live now?

The most plausible rival to the broad conception is the first of the narrow conceptions given in the last paragraph: confine the perspectives represented to those of nation-states. Consideration of other ways of narrowing poses a challenge to all the rivals. What is it that makes narrowing in a specific way acceptable and further confinement illegitimate? Why not the gentlemen, or the members of Mensa, or the tycoons, or the scientists? The best answer focuses on the status of nations as economic units. Nations produce the resources needed for supporting inquiry, and that gives citizens of a particular nation a privileged voice in determining scientific significance. If Americans are contributing more to supporting Science, their needs should be given greater weight.

Seductive though it may initially seem, this line of argument is obviously dangerous. If division by productivity is appropriate, why not carry it through more finely and on a broader range of issues? Strictly speaking, the productivity of a nation emerges from the efforts of its individual members. Why not give those who generate more a proportionally greater say? Why not apply the principle across other decisions, apportioning votes on electoral offices according to the contributions made to the national resources: to each according to his productivity . . . ? Although questions of this sort embarrass the narrow conception, the more basic reasons for its unacceptability lie in the general approach to value found in chapter 2.

Imagine that ideal discussion according to the broad conception would license a line of inquiry that would be rejected on the narrow conception. Suppose further that the pertinent research program has severe consequences for some particular population of nonaffluent people. How could the decision not to pursue it be explained to them? It would have to be acknowledged that their perspective was not included in an ideal deliberation, and that the basis for leaving them out was economic: the evaluation of significance proceeded by consulting those who contributed the resources to be used. If all nations had resources available for supporting research meeting their particular needs, the project would not falter, since the group to whom it is important could include it on their own agenda. Because of their poverty, however, they are in no position to pursue scientific research. Hence, proceeding on the basis of the narrow conception has severe consequences for them.

Behind the contemporary worldwide distribution of resources stands a long and tangled history. It would be very hard to defend the judgment that all are now rewarded according to their deserts: the route to the present has involved all sorts of murky acts, as well as plenty of luck. *There is thus no basis for any group of people with a heterogeneous distribution of resources to accept the view that decisions profoundly affecting human welfare should be made on the basis of restricting views to those who happen to have done well.* As a result, even a generic commitment to the approach to values in terms of discussion under conditions of mutual engagement, one that does not yet specify the size of the population of discussants, cannot endorse the grounds on which the narrow conception is based. The framework of chapter 2 thus requires the broad conception for the ideal of well-ordered science.

19. WELL-ORDERED SCIENCE: DEFENSE

Many people, especially scientists, react to a plea for democracy with alarm, insisting on the autonomy of scientific practice. Part of the fear stems from suspicions that democratization, even in the guise of well-ordered science, will submit research to the tyranny of ignorance. It is worth repeating that well-ordered science is deliberately designed to overcome this problem, that it imposes stringent cognitive conditions, and that it assigns an important role to the authority of experts. Moreover, scientific autonomy, like that of other agents, covers many spheres of activity, and it is important to understand just which of these might be threatened. Very likely, the image of the autonomous scientist is a residue of the original commitment to private activity, embodied in the "gentlemen's clubs" of the early modern period, no longer apt when Science has become central to the public knowledge system—despite Vannevar Bush's ingenious attempt to combine public support with the maintenance of a class of Platonic guardians.

In this section, I plan to respond to some common objections to well-ordered science, many of which descend from insistence on scientific autonomy. The most basic form points out that the scientific community has a clearer vision of the collective good to be realized through inquiry. As things stand, that is probably correct—and if one had to pick any single group to decide what lines of investigation to pursue, scientists would be the most appropriate choice. Yet the asymmetry between scientists and the lay public should not be overblown. One of the most fundamental thoughts behind democracy is that individual people have a better understanding of aspects of their own predicament than do outsiders, however wise and well-intentioned. Some years ago, a team of investigators visited a group of

African pastoralists and discussed with them the possibility of developing vaccines for their children. Their interlocutors asked for some time to ponder the issue and, when they returned, made the unexpected suggestion that a vaccine for their goats would be even more welcome.[7]

Furthermore, as anyone who has ever heard different groups of scientists debating the promise of their own special fields will know, even if the scientific view is more farsighted than that of outsiders, it is typically myopic. Each specialist tends to view the scientific universe in the style of Saul Steinberg's famous Manhattan cartoon (in which the Upper West Side has far greater prominence than Middle America, the Pacific Coast, or Asia). To construct any balanced view of research possibilities would require something like the ideal conversation envisaged, at least among representatives of various scientific fields, and, when the insights of individuals into their own needs are appreciated, it becomes evident that outsiders ought to be included. Well-ordered science emphasizes the importance of tutoring, precisely because, to pursue their interests, the outsiders will need the various kinds of special knowledge the scientific community can supply. Rather than trying to drown out responsible judgments with a chorus of ignorant voices, its conditions fuse the different kinds of knowledge distributed through the human population.[8]

Champions of autonomy will proceed to more sophisticated objections. "We already know," they declare, "that directed scientific research goes badly; that it has been wonderfully fruitful in the past for brilliant scientists to explore their hunches, that unanticipated benefits come from inquiries into apparently impractical questions, and that the course of science is unpredictable." Arguments like these are often made from the armchair—or *ex cathedra*. The autonomist has a few bits of anecdotal evidence, having read a book on Lysenkoism and a biography of Einstein. In fact, little is known in any systematic way about the responsiveness of scientific research to social directives. The basis for any hypothesis about the bad effects of something like well-ordered science is extraordinarily thin. The autonomist's pronouncement rests on the sorts of judgments routinely denounced in basic courses in methodology in any scientific field: sketchy histories are invoked without any attention to sampling or to proper comparisons. So far, the social study of scientific knowledge cannot deliver a statistical basis from which anyone can project the likely effects of attempts to plan different kinds of research. More fundamentally, however, insofar as genuine knowledge about social direction of inquiry, success of brilliant individuals, or fruits of research into pure topics, is, or becomes, available, that knowledge could and should be employed to further the democratic process. *It should be part of what the ideal discussants know.* So, in their deliberations, they can take into account the track record of different attempts to direct inquiry.

Many contemporary molecular biologists would frown on centrally directed ventures that attack prominent medical problems—analogs of the "War on Cancer"—insisting that the route to success is often indirect. Their reservations are not antipathetic to the ideal of well-ordered science, however. To acknowledge a particular problem as practically significant, as when cancer is seen as requiring major scientific effort, is not to favor any specific strategy for addressing that problem; for example, a blind assault that dismisses all attention to "basic issues" in pertinent sciences. Strategy should be informed by what is known about the past successes and failures of various ways of conducting investigations aimed at similar ends—and that is exactly what well-ordered science demands.

The last part of the autonomist's protest deserves a slightly different response. What exactly follows from the fact that we cannot foresee the course of science? Is it supposed that no decision we can now make about issues that matter is preferable to any other? Are past attempts to allocate priorities among lines of scientific research arbitrary and capricious? If so, the autonomist's own confidence in the wisdom of scientific judgments would be undermined. You might just as well toss coins or read tea leaves. The practices the autonomist wants to preserve testify to our understanding that, while we cannot make fine-grained predictions about what research will bring, we are not completely clueless. We know for example that needs are more likely to be met if more effort is expended in one direction rather than another: stepping up research into mechanisms of gene transcription is not likely to enable us to slow global warming—it might, but the probability is not high. The scientific situation is, again, akin to ordinary circumstances of decision making. Families plan for the education of children and the retirement of parents in ignorance of crucial information about what the future will bring. They know that unforeseen contingencies might disrupt the most well-considered plans. Responsible people do not conclude that they might just as well spend to their credit limit (or beyond). In light of the best judgments they can make, acknowledged as rough, they seek the most likely paths to achieve their ends. Working in concert, the scientific community and the broader public ought to be able to achieve something similar.

It may still appear, however, that the ideal of well-ordered science is too pragmatic, too restrictive. Should there not be some place for people whose lives are centered on projects of disinterested inquiry, whose plan for their mortal span consists in answering questions about which others do not care? They do no harm. Often they work for no great rewards. Why should the Republic of Letters not allow them a modest place?[9]

Even under well-ordered science there might be room for impractical dreamers—for there might be benefits to all from allowing them to follow their own fancies. During the Early Modern Period the status of mathematicians changed, as it became evident that new extensions of mathematical language might prove useful resources for inquiry generally. In effect, mathematicians were given a license to address esoteric questions they found interesting, and the decision to grant that license has paid off handsomely. So, too, it might be more generally. In the end, however, making a place for the satisfaction of refined curiosity ought to be defensible in the ideal democratic conversation well-ordered science envisages. The informed deliberators ought to be able to recognize the value of pursuing these inquiries, even if the benefits are indirect. They ought to be convinced not only that no harm is done but also that the talents of these investigators are properly used, contributing to the broader human good. To think otherwise is to yearn for the existence of seventeenth-century gentlemen "free and unconfin'd," even though the social world has changed and Science has become central to public knowledge. That change brings responsibilities that those drawn to "projects of pure disinterested inquiry" ought to recognize.

Turn now to a different worry about well-ordered science, not that it constitutes a clumsy form of interference with a valuable institution (one better left to the wise people who contribute to it) but that it does too little to change the status quo. Is the ideal toothless? Can it be developed precisely enough to recommend modifications in current research agendas?

Consider contemporary biomedical research. Most of it is carried out in affluent societies, and almost all of it concentrates on diseases afflicting people in those societies. (At least, that is what the community of researchers tells those who ask what they are doing; a closer look would surely reveal many investigators working on "pure" problems in "basic biology," questions whose significance they could usually defend as likely to yield medical advances in the more or less distant future.) Contrast the distribution of disease research with the statistical data on worldwide disease and disability. Diseases that cause a vast amount of human suffering, particularly among children, receive only a tiny part of the investigative effort. In some instances, that is because the pertinent disease has already been "solved": a method of prevention, cure, or treatment is available and can protect children in the affluent world. The fact that the method cannot be imported into the circumstances in which poor children live—and fall ill, and die—does not affect the status of the "solution"; it is not, on that account, recognized as partial.

Well-ordered science recommends a plausible principle: the *fair-share principle*. Waiving considerations of tractability, each disease should be investigated according to its contribution to the total suffering caused by disease. A simple measure, applicable only to fatal diseases, would measure the contributions by the numbers of resultant deaths. More subtle appraisals discount the years of a person's life by the disabilities to which she is subject. However the contributions are assessed, if the principle is applied directly to the statistics on disease incidence, it is evident that actual research into diseases is skewed toward conditions affecting affluent people. Many diseases that kill or incapacitate poor people receive support on the order of one-hundredth of their fair share (Flory and Kitcher 2004; Reiss and Kitcher 2009).

Mechanical application of the fair-share principle would be foolish, since considerations about profitable inquiry should attend to considerations of research promise. Hence, the formulation given introduced the proviso that issues of relative tractability were waived. Consequently, the actual distribution of research effort might be defended by proposing that the affluent diseases actually investigated—possibly even overstudied—are especially likely to yield important insights. Any defense along these lines would have to cope with the fact that contemporary biomedicine supplies promising tools for tackling diseases that bring misery to millions. Genomic sequencing of pathogens offers clues for designing effective vaccines capable of transportation to the environments in which they are needed. There are no sure-fire strategies (particularly in the case of rapidly mutating infectious agents), but knowledge of the genome can indicate potential genes, encoding proteins likely to appear on the surface of the disease vector; if such proteins can be inserted into benign micro-organisms, it is possible for them to produce antibodies to the pathogen. In contrast to many diseases currently attracting large support (because they afflict rich people), a large number of understudied diseases of the poor lend themselves to a more systematic program of research. If anything, these diseases are *more* tractable than those actually investigated.

Well-ordered science requires refiguring of the medical research agenda in light of considerations of global health. Even without articulating the ideal further than I have done, it is possible to recognize places at which our actual practice would be revised. That is because the very basic needs of many people are not met, and because there are lines of inquiry promising to relieve this situation. Surely there are many instances in which it would be hard to predict much about the outcome of a conversation under conditions of mutual engagement, yet problems bearing on the health of children in regions of high mortality and disability are not among them. However they are tutored, deliberators who represent those children and their parents will be expected to continue to feel, and to express, the pains those children and their families experience. The details of the ideal conversation need not concern us when one feature of it is so evident.

Contrast this example with a different question, one that has figured in earlier sections. How is the balance between "pure," or "basic," research and investigations directed toward immediate problem solving to be struck? Recall a conclusion of §17: there are two potential bases for justifying attention to "pure" questions: pursuing them is likely to produce tools for solving a wide range of practically significant problems down the road (Vannevar Bush's "seed-corn" argument); answering them would satisfy widespread human curiosity. How these lines of justification play out in ideal deliberation depends on crucial details about the state of the sciences and about the needs of contemporary and future people. Some problems requiring investigation may be so urgent that counsel to wait for the fruits of "basic" research would ring hollow. In other areas, ideal deliberators might judge either that direct attempts, undertaken without more "basic" understanding, would be futile, or that stopgap measures could be deployed while the research community sought a more systematic solution.

Without extensive further information about the research opportunities now available—the sort of information that would be provided in the tutoring well-ordered science envisages—it is impossible to be certain that "pure" questions, conceived as stepping-stones to future practical benefits, would inevitably figure in the agenda of well-ordered science. Perhaps aggregate human needs are so urgent that we should deploy the knowledge already gained to craft directed programs of inquiry to satisfy those needs as speedily as possible. Precisely because the knowledge available is, in some areas, so powerful, so susceptible of further development, and precisely because it has often grown out of programs of "basic" research, it appears highly unlikely that the ideal deliberation would abandon so profitable a historical strategy. Should we not imitate those many scientists of the past who posed and answered questions without any obvious pragmatic payoff—the physicists, chemists, and biologists whose basic research underlies countless present technologies? Unless you suppose the situation is truly critical, that our species faces practical problems that command direct attention, well-ordered science is likely to maintain a role for "basic" research.

Is that enough? Section 17 assigned a place to the satisfaction of human curiosity, independent of any practical benefit. Achieving satisfaction of that sort is valuable *in principle*, but it does not follow that ideal deliberators would be moved by it. Here, the outcome of the ideal deliberation is even less certain. Without a far more detailed survey of aggregate human needs, of the possibilities of addressing them directly, and of the theoretical projects justifiable on the basis of their promise for future strategies of intervention, nobody can predict how the ideal conversation would come to conclusion. Can you rule out the scenario in which research directed at immediate relief, together with lines of "pure" research that promise future fruits in application as well as answers to questions that arouse curiosity, are so abundant and so compelling that they leave no place for the luxury of knowing something "merely" for its own sake? To allow, as I have done, that disinterested "pure" understanding has a value, that it should be placed in one balance of the scale, does not guarantee its being sufficiently weighty to offset whatever occupies the opposite pan. Genuine doubt is appropriate here.

Scientists, especially those fascinated by the aspects of nature they study, will probably find this conclusion troubling—even grounds for doubting the ideal of well-ordered science. They should not. For their own perspective is registered in the ideal deliberation, their own fascination with (say) the hominid family tree is conveyed to their fellow discussants, who feel its force as they do. On what basis could they object if, after serious sympathetic engagement with all human perspectives, the practical needs of others seemed more urgent? Could they themselves engage with the other perspectives, think themselves into situations in which more elemental things than the satisfaction of curiosity are lacking, and still insist that their own "pure" questions merit attention? Especially if, as I have conceded, the search for "basic" understanding will continue in areas in which it is coupled to future practical promise.

As we shall see in chapter 7, the value of satisfying curiosity is one—like freedom (§11)—deserving attention to its distribution. Defending the value of "pure" knowledge for its own sake is far easier if the benefits of refined understanding are widely available. Placing private satisfactions of this sort ahead of attention to urgent human problems of an elementary sort involves a failure of altruism—and remedying altruism failures is the original function of the ethical project (§6).

20. MERELY AN IDEAL?

Well-ordered science is an *ideal.* It may seem a utopian fantasy, the sort of thing that may figure in philosophical discussions but that has little place in a realistic account of the sciences (Lewontin 2002). There is an important distinction between specifying an ideal, something at which our practices should aim, and identifying procedures for attaining or approximating the ideal. To proceed to the latter task requires a large amount of empirical information, information no one yet has. Nonetheless, meaningful ideals are those for which we can envisage a path that might lead us toward them, and a philosopher who proposes an ideal should be able to point to the initial steps we might take (as Dewey insisted; it is also important to appreciate that, as we move toward an ideal, our conception of it may be refined).

Actual deliberations about the ends of the sciences are often, probably always, infected by special interests, ideological presuppositions, and inequalities of power. These facts do not diminish the importance of the ideal. They suggest difficulties to be overcome in realizing the ideal, ways in which well-entrenched features of political life might need amendment. To scoff at philosophical ideals on grounds that they require a lot of changes would be a serious mistake, for, without some understanding of where you want to go, efforts to improve on the status quo will be leaps in the dark.

My attempts to identify some steps forward will begin from diagnoses of respects in which the current framing of investigation departs in striking ways from well-ordered science. I offer four hypotheses that develop points made earlier in this chapter.

1. Present competition among scientists and fields of science is con-

strained by historical contingencies that no longer reflect human needs. Even if you thought scientists were the only people whose judgments should count in setting the research agenda, you ought to worry about the ways in which priorities are set. As already remarked, individual scientific visions are parochial. They are often pitted against one another in an arena in which there is no serious possibility of surveying the merits of competing possibilities, and in which institutional structures partition potential research proposals into areas defined in the past. History frames the current distribution of public research, often in baroque fashion. For example, the current ramshackle organization of the National Institutes of Health reflects the accidents of the past.

2. *The flaws of vulgar democracy are inherited by existing systems of public input.* Vulgar democracy is problematic because the preferences expressed are untutored. Contemporary public procedures for shaping the research agenda proceed from two sources: either government (typically responses to large perceived problems but often slanted toward constituencies deemed important by the politicians involved) or special groups of concerned citizens, sometimes well-informed about the issues they raise (local pollution, say, or a particular disease) but ignorant about the full range of scientific possibilities and the diverse needs of their fellow citizens, let alone those of more distant people. Priorities are set as a result of haphazard shouting of more or less powerful voices, each expressing, at best, some partial truth. Possibly public input of this sort improves the results that would be achieved if the scientific community were left to its own devices—that is an empirical issue about which we have little evidence—but there is no reason to think it takes us far toward well-ordered science. Insofar as we introduce democracy into thinking about Science, we incorporate the adversarial rather than the deliberative elements.

3. *Privatization of scientific research will probably make matters worse.* Government pressures and the clamor of interest groups sometimes have the advantage of representing people with urgent needs. Private investment in scientific research, ever more apparent in biomedical investigations and in the information sciences—the two fastest-growing fields of inquiry of our age—is, in both the long run and the short, tied to considerations of financial profit. One immediate result, of concern to many biological researchers, is the neglect of "basic" questions in favor of areas in which profits can be expected. The decisions issuing from two large groups, the scientific com-

munity and the general public, are likely to be dominated by a clash of parochial visions. Nevertheless, each of these groups has some connection with the ideal deliberators of well-ordered science: the scientists appreciate the significance of achievements in their own specialized areas, members of the public recognize their own urgent needs. Entrepreneurs are at a further remove. To the extent their decisions respond to genuine needs, those needs will be raw and untutored, typically self-directed, *and they will be the needs of those who pay.* Markets sometimes work the wonders frequently attributed to them, but there are systematic reasons for thinking that, in shaping scientific research, an unregulated market will produce a travesty of well-ordered science.

4. Current scientific research neglects the interests of a vast number of people, except insofar as their interests coincide with those of people in the affluent world. The example of the distribution of biomedical research and the deviation from the fair-share principle provides a striking illustration of a potentially general phenomenon. The world's poor are only accidentally represented in decisions about the lines of inquiry to be pursued. Without a more detailed understanding of their needs and aspirations, it is impossible to know just how much difference this makes, that is how frequently the neglect manifested in the biomedical case obtains.

Although each of the hypotheses is plausible in light of obvious features of our current situation, more detailed information about the attributed effects would be welcome. Assuming the diagnoses are roughly correct, it is not hard to envisage steps toward well-ordered science. I offer some pervasive problems and proposals.

Myopia. Even when informed and well-intentioned scientists try to think broadly about research options, their discussions suffer from the absence of a synthetic vision. Instead of pitting one partial perspective against another, it would be preferable to create a space in which the entire range of our inquiries could be soberly appraised. We would do well to have an institution for the construction and constant revision of an *atlas of scientific significance.* That atlas would provide maps of the various fields of inquiry, showing how significance accrues to the work that has already been done and how it might be extended in significant ways. It would connect the technical work of specialists with broad issues about which people are curious and with practical consequences for human lives. The resultant maps—*significance graphs* (Kitcher 2001)—would enable everyone, scientists and the public alike, to appreciate the full range of current opportunities, to understand all the ways in which some inquiries might, given our best present judgments, bear fruit. The atlas would allow a more reflective view to replace the competing myopic visions offered by (understandably enthusiastic) specialists.

Ignorance of science. The atlas is part, though by no means the whole, of what is required if public input into science policy is to come closer to well-ordered science. Central to democracy is the thought that people can take political action to express their *interests*, not merely the variously misguided preferences they might have. Even before we envisage deliberators who are sensitive to the interests of others, it is important that their self-directed wishes be enlightened. Many people around the world oppose measures intended to develop alternative forms of energy and strongly want to continue their familiar practices of fuel consumption. Most of these people have a far deeper and more central wish that the world in which their children and grandchildren live should be habitable, not subject to violent disruptions that would create massive difficulties in obtaining shelter, food, water, and protection against disease. According to the contemporary consensus in climate science, these people's desires are in a state of considerable internal tension: policies framed in accordance with the short-term wishes (energy consumption as usual) threaten the more central wish that descendants will thrive. The case of climate policy is one of the starkest instances of unidentified oppression, but widespread ignorance of important parts of public knowledge contributes to many gaps between the preferences citizens express and their most central interests. If public input into scientific research is to overcome the perils of vulgar democracy, steps must be taken to increase levels of scientific literacy.

How might this be achieved? The problem is many-sided, and we shall be considering aspects of it in later chapters. For the moment, focusing on the possibility of steps toward well-ordered science, two ways of improving communication between Science and the public deserve consideration. The first would proceed from the sciences out toward the public. In recent years, there has been a shift in attitude within the scientific community, a sense that spokespeople for major scientific fields are valuable, not to be dismissed as vulgar "popularizers" or reputation-seeking "has-beens." Writers like Carl Sagan, Stephen Jay Gould, E. O. Wilson, Richard Dawkins, and Brian Greene have done valuable service by explaining major ideas lucidly and elegantly. Their writings and their television appearances have greatly expanded public understanding of science—and it was a signal of their achievement that the United Kingdom instituted professorships in the public understanding of science and appointed Dawkins to the first chair (at Oxford). This trend could be extended far more widely, and scientists who are especially good at communication could be encouraged to view this as a central part of their mission.

Conversely, it would be possible to create groups of citizen representatives, drawn from diverse segments of different societies, who would undergo some practicable approximation to the tutoring envisaged in well-ordered science. These people would be "led behind the scenes," brought into thriving areas of scientific research, and given explanations of the state of knowledge, of the lines of envisaged future progress, and of the accompanying difficulties, as the specialists see these things. The atlas of scientific significance would be explained to them. After discussions with one another, they would then be available to the broader public, to report on their—nonexpert but informed—understanding of the state of inquiry, and to discuss possibilities with a much wider group of lay citizens. In light of these discussions, they could then return to conversation with specialists, acting as intermediaries in facilitating information flow and dialogue.[10]

Ignorance of others. The ideal deliberators envisaged by well-ordered science not only understand the state of scientific knowledge but also recognize one another's needs. Although no readily constructible institution could provide all the nuanced understanding available in the ideal conversation, it is surely possible to remedy some of our ignorance. The atlas of scientific significance could be supplemented with an *index of human needs*. That index would be built up by systematically exploring human problems as they are perceived by the people who encounter them. Ideally, the investigations would proceed by striving to isolate deep desires, real interests that might sometimes be masked by distorting ignorance, so that here, too, there would be efforts at tutoring to clear away common misapprehensions and problems of unidentifiable oppression. Imperfect though such efforts would be, even rough approximations would enable research to be guided in ways that no longer leave out large segments of our species.

Failure of sympathy. Ideal conversationalists not only know the wishes of their fellows, they also adjust their own preferences to accommodate others. Overcoming ignorance about the plight of people whom inquiry often neglects should be the prelude to sympathetic identification with them. Here, as in the case of ignorance about science, the issue is many-sided. One part of a remedy would take seriously the idea that part of education consists in the encouragement and expansion of altruistic tendencies.

Another would lie in commitment to exposing cases in which scientific research is distorted through subordinating benefits for many to economic profits for the few. Scholarly research sometimes reveals how inquiry is directed toward ends quite different from any public good: pharmaceutical companies do not produce a drug that could cure thousands of poor children because there is no profit in it; well-known scientists with ideological commitments or ties to particular industries block public awareness of important information (Oreskes and Conway 2010). Commentators on the sciences need to pursue inquiries of this sort more widely, and their findings, when well documented, should be well publicized (that is part of the responsibility of journalism). As the public information system becomes fractured between public and private forms of support, it is important to keep track of the places in which an "invisible hand" really does operate, producing outcomes that yield widespread benefit and those in which the market harms the many for the enrichment of a very few.

All the proposals I have made need further refinement and development. They are attempts to respond to the challenge posed at the beginning of this section, to show that, even though well-ordered science imposes strong—unrealistic—conditions, we can nevertheless envisage steps to take us closer to it. In specifying the path more exactly, it is possible to learn through small-scale social experimentation. Researchers can investigate—and have investigated—the merits of various ways of improving communication among different groups or facilitating outside oversight of decision making (Fishkin 2009; Jefferson Project). The institutions whose functions I have sketched would best be fashioned in light of such research and through trial of various possibilities. In proposing that we explore in this way, I reiterate a theme of earlier discussions: our system of public knowledge is the product of a tortuous history, and there is little reason to think it has delivered to us a set of institutions insusceptible of any improvement.

21. CONSTRAINTS ON PURSUIT

To close my discussion of questions about investigation, it is worth looking more briefly at the phase of inquiry that follows the setting of the agenda. Various important questions have been isolated (and we hope the decisions correspond approximately to those that would have occurred under well-ordered science). How are they to be pursued?

Typically, we expect investigators to be well-informed about the best methods for achieving their goals, and to follow those methods. We shall later consider complications that might arise when there are various possibilities for proceeding (chapter 8). For the moment, however, attention will be restricted to two main issues: Who are these investigators? Are they subject to constraints that might not be commonly appreciated?

During recent decades, it has become a commonplace that certain ways of doing research are not to be tolerated: commentators look back in horror on the notorious Tuskegee experiments (in which African Americans known to be infected with syphilis were deliberately left untreated) and on the "science" undertaken by Nazi doctors in the death camps. Ethical limits are imposed, even when the cost of the restrictions is that questions we hope to address become more difficult or even unanswerable. Sometimes, urgent issues about the causes of a disease could be settled by selectively exposing people to pathogens; we could answer questions about nature and nurture relatively directly by separating carefully selected children from their families and rearing some of them in bizarre environments. All scientific communities now acknowledge ethical constraints forbidding such experiments. Communities also frown on scientific piracy, attempts to acquire without consent the data obtained by others, although they also recognize that scientists have obligations to share their findings. Do these restrictions exhaust the proper constraints on the pursuit of knowledge?

The examples just given have more subtle relatives. Much contemporary research employs sentient animals, sometimes submitting them to unusual pain, sometimes bringing into existence creatures whose lives will be short and unpleasant. A total ban on experiments that inflict suffering on animals would inhibit many lines of inquiry with great potential for alleviating human agony and misery. A completely tolerant attitude toward animal suffering would allow investigations that pursue trivial goals. Where are the lines to be drawn?

The perspective on value-judgments of chapter 2 provides a basis for decision. Again, it is a matter of balancing valuable goals against one another. As in the framework of well-ordered science, a proper verdict would be one achieved by ideal deliberators, well tutored and mutually engaged, who considered the effects both of the proposed experiments and of not undertaking them. Since the affected parties are people who suffer from disease, on the one hand, and sentient nonhuman animals, on the other, it is crucial that both these groups be represented in the conversation.

How can that be? Even if a tiny few of the animals who would be affected are credited with some linguistic skills, those are far too rudimentary for them to engage in the kind of conversation envisaged. Indeed, there appears to be a very general objection to my approach to values; to wit, that it arbitrarily excludes our many sentient relatives. Despite its inclusiveness with respect to the human population, is it guilty of an illegitimate human chauvinism, something some might take to be as noxious as ethical stances that have confined their attention to a small subset of the human population?

Ideal conversation already has to represent *people* who cannot speak for themselves. Members of future generations are not available to comment, nor are the very young, nor are adults suffering various types of disabilities. Their perspectives are to be included through representation by involving people who know them intimately and who are devoted to their interests. So, too, for nonhuman animals. If we were to try to simulate an ideal conversation about the propriety of using an animal subject in a particular fashion, it would be important to involve people who could supply details about the animal's physiology, its responses to various kinds of pain and deprivation, its kinship with human sufferers, and so forth. Equally, it would be crucial to include those intimately familiar with the sufferings of human disease victims, people who might obtain relief if the animal experiments were permitted.

The situation might well be completely symmetrical. Some forms of human disease strike suddenly and, when they do, preclude any possibility of the patient's testifying on her own behalf. Under such circumstances, both primarily affected groups have to be represented by others, and the responsibility of the representatives is to provide an adequate account of the consequences, one that will enable everyone to reach a decision about whether to allow the proposed program or to debar it. No doubt these choices will sometimes be hard—although actual cases are often more tractable than the stark scenarios beloved of abstract philosophy, in that there are ways of minimizing animal suffering or pursuing human benefits along alternative lines. There is, I submit, no better way to make them than to replicate, to the extent we can, a conversation that proceeds through mutual engagement with all the potentially affected parties.

It is similar for other difficult examples. Sometimes people who are passionately dedicated to particular causes, or who know that their lives will soon end, volunteer as subjects in experiments in which involuntary participation would be banned. If those people truly feel that participation is a constituent of their life project, central to who they are and what they aspire to, preventing their noble sacrifices would be an interference with their freedom. The obvious suspicion is that some form of coercion has been at work, that, at bottom, these people are no more expressing an autonomous choice than were the Tuskegee subjects or the "patients" of the Nazi doctors. To address worries of that sort, we can adapt the ideal of well-ordered science and the more general approach to value-judgments in terms of ideal conversation it embodies. Volunteers would discuss their plan of action with people of different perspectives, including some who were dedicated to their welfare and some who were suspicious about social coercion, aiming to replicate insofar as they could conditions of engagement with the would-be experimental subjects. Permission would depend on their final agreement.

Or consider more complicated cases of scientific piracy. We frown on stealing data when those with important information act swiftly to release it to their colleagues. There are, however, envisageable cases (maybe actual examples) in which the data are urgently needed and a pathologically doubtful investigator feels the need for further trials. Does a genuine ethical constraint debar someone deeply concerned with people threatened by the delay from attempting to tease the findings out of subordinates in the lab or even to find ways of looking at a crucial photograph? Is the dithering doubter living up to the proper responsibilities of a scientist? To answer such questions, we can only appeal to judgments about the case, formed through the best approximation to ideal deliberation we can contrive.

Contemporary practice, especially in the biomedical sciences, already embodies a healthy approximation to the ideal and even the procedures I am recommending. Although researchers sometimes complain about them, institutional review boards provide good ways of elaborating and applying constraints on research. If their discussions are currently problematic, that is not because of the recourse to conversation but a result of the channels through which the discussion flows. On the approach to values recommended in chapter 2, decisions should not be made by wielding abstract principles (of the sort medical specialists absorb from simple philosophical textbooks and struggle to apply) but through deep immersion in the case from a variety of human perspectives. As the cognitive conditions on mutual engagement demand, the conversation must not be stopped by announcing religious precepts—for those have dubious authority—but neither is there any secular source that can transcend the authority of the conversation. Promising practices of research review could be improved by finding ways to bring deliberation closer to conditions of mutual engagement (perhaps by increasing the diversity of perspectives), and, in principle, those practices could be applied elsewhere, as in the imagined conflict between the dithering doubter and the researcher who desperately needs the results currently withheld.

The example of possibly permissible piracy introduces a point about scientific responsibility, for we might judge that the doubter fails to live up to the demands of responsible investigation. Coordinated activities require people to do their bit, to discharge the tasks assigned them so that a common goal can be realized. Thinking of Science in this way, as a collective attempt to expand and refine public knowledge, imposes apparent burdens on researchers—they are no longer "free and unconfin'd." They can be held accountable not only for what they do but for what they fail to contribute. We can approach my other question about the pursuit of knowledge—Who are the investigators?—in light of this perspective.

Consider an obvious extension of well-ordered science. At the end of the ideal deliberation through which they have drawn up the agenda, the discussants turn to a different question. How are they to assign the members of the community of researchers to the projects they have selected for pursuit? We can imagine them to have extensive knowledge of track records and talents. Combining this with the priorities they have set, they act as field marshals, assigning the troops to their tasks so as to maximize expected success.

Confronted with this totalitarian vision, many people—perhaps all scientists—will surely protest. Here is an intolerable invasion of autonomy! No scientist should be told what research project to undertake! These protests are entirely justified. The imagined extension of well-ordered science is compatible neither with democratic ideals nor with the approach to values I have proposed.

To see why this is so, it is useful to distinguish two different questions:

(A) Are scientists ethically required to undertake the kind of work that would best advance the community goal (the promotion of public knowledge in the form emerging, in the context, from the ideal deliberation)? (B) Should there be a procedure within Science for making scientists do what is ethically required of them (assigning them to the tasks that would best advance the community goal, or punishing them if they refuse to undertake those tasks)? Notice first that, even if you were to believe in an affirmative answer to (A), you might give a negative answer to (B). There are many kinds of human conduct that depart from or violate ethical obligations that we do not bring within the scope of coercion or punishment—and for which we think that coercion or punishment would be a breach of ideals of freedom. Democracies rightly leave leeway in the ethical choices of citizens.

It is wrong to suppose, however, that the answer to (A) is an automatic yes. The supposition descends from a misunderstanding of the attitudes of the ideal deliberators. They are imagined as having finished setting the agenda and proceeding to the optimal distribution of tools for its implementation. *Were they to proceed in that way, they would be guilty of a crass failure of mutual engagement.* For the scientists they envisage "assigning" are not *tools* but *people* whose perspectives and projects ought to be represented in the deliberation. Under many circumstances, the fact that scientist X is already passionately committed to thinking about question Q makes X less good as a candidate for inquiring about Q*, even though, without that passionate commitment, X would be the optimal person to investigate Q*. Furthermore, even when X's passion for Q would not invalidate his status as the best investigator of Q*, that passion should be taken into account and respected—for the ideal deliberators recognize its role within X's life projects, and they are concerned, when other things are equal (or approximately equal) to promote success in life projects. The perspective on values adopted in chapter 2 thus allows for cases, probably the overwhelming majority, in which scientists have no ethical obligation to pursue questions other than those they freely and reflectively choose.

There will, however, be occasions on which the ideal conversation is more demanding—*and these correspond to obligations we already recognize*. Suppose that Q* is enormously important, that the lives and projects of many people turn on investigating it successfully. Assume further that X is significantly more likely to carry out the investigation successfully than anybody else, and that X, like everyone else, knows that. Although Q fascinates

X, an answer to it is not particularly urgent. Under these circumstances, ideal deliberators would conclude that X has an ethical obligation to pursue Q*—and if X were to engage in conversation under conditions of mutual engagement, X would appreciate the obligation. Although they are not common, circumstances of this sort are quite familiar. A state of emergency calls suddenly for particular lines of research, and scientists drop what they have been doing and play the roles others ask of them. They go, for example, to Bletchley or Los Alamos.

Fear of well-ordered science as leading to the research gulag is unfounded. It is possible, however, that ideal conversation, aimed at balancing the claims of private projects and the public good, might impose more general obligations on people—not just researchers but all citizens who contribute to some collective enterprise—than those of which we are currently aware. We should come to view those obligations as a broadening of the special instances we already recognize, when, say, some large danger calls for us to modify our activities, as a call to greater unselfishness.

The distribution of researchers reflects individual preferences, but it is entirely legitimate for the community to offer incentives to guide investigative effort toward important projects, currently neglected. (Chapter 8 will take up this issue more systematically.) I want to close with a brief look at a related question. It is tempting to think of the pursuit of knowledge as a closed enterprise: promising young people are thoroughly trained, and, eventually, they become part of a community whose members address the questions viewed as significant. Outsiders are not expected to make any contribution. Possibly their efforts are even discouraged.

There are some areas of inquiry in which efforts to bypass the standard training regimes waste time and resources. To address the technical questions adequately requires specialized knowledge, to operate the equipment properly demands experience. Anyone who has ever edited a professional journal is familiar with the submissions that confidently claim to overturn received knowledge: the "refutations of Einstein," for example. It is not always so, however. G. H. Hardy deserves enormous credit for his willingness to read far enough in the curious letter sent to him from India to recognize genius, even if that genius was expressed in unfamiliar, even amateurish, terms.

Even in a predemocratic society, scientific research was open to the contributions of outsiders. The gentlemen of the Royal Society listened to the reports of sea captains who had visited distant regions of the globe (even if their accounts of mermaids in the Sargasso Sea were not altogether reliable). Democratic societies might well explore ways of making greater use of people who are not professional scientists: naturalists with an eye for local flora, dedicated amateur astronomers. As I finish this chapter, news sources have announced some interesting results achieved by an unusual team of investigators. Computer game aficionados have made some advances on the recalcitrant problem of protein folding, not because they have deep chemical knowledge but through their experience in transforming images on the screen. Thanks to the design of an ingenious game—Foldit—a different set of skills can be marshaled for scientific inquiry. It provides an interesting precedent for further ways of widening the set of contributors to public knowledge and thus not only advancing a specialized field but also making Science more democratic.

Chapter 6

PUBLIC REASON

22. PROBLEMS OF CERTIFICATION?

I now turn to a far more disturbing issue: whether there is a role for non-scientists in the certification of scientific claims. This chapter will argue that there is, albeit not of the simple and radical kind that (rightly) perturbs scientists.

Proposing that democratic values should play a role in decisions about what lines of investigation should be *pursued* seems a relatively conservative move, although it is worth recognizing that those decisions shape the future body of public knowledge that serves as background to the assessment of subsequent claims. Well-ordered science thus has indirect effects on processes of certification—for the simple reason that *any* way of framing the context of investigation would do so. How new candidates for knowledge are evaluated depends on what is currently known. There is no avoiding dependence on the contingencies of history.

A more radical idea, most famously proposed by Paul Feyerabend (1978), would introduce democracy directly into processes of certification. According to Feyerabend, "it would not only be foolish *but downright irresponsible* to accept the judgment of scientists and physicians without further examination. If the matter is important, either to a small group or to society as a whole, *then this judgment must be subjected to the most painstaking scrutiny.* Duly elected committees of laymen must examine whether the theory of evolution is really as well established as biologists want us to believe, whether being established in their sense settles the matter, and whether it should replace other views in schools" (1978, 96). When scientists and philosophers of science emphasize the autonomy of inquiry, their fears of "mob rule" react to exactly the "freedom" Feyerabend wants to applaud.

It is not hard to understand how Feyerabend's argument works. Suppose you were to doubt the existence of any method for resolving major scientific disputes. You would view the transformations of our ideas about the natural world out of which our present perspective has grown as a series of decisions that might, with equal reason, have gone in other ways. Each of those decisions embodies the values of the victorious party, a majority of those embroiled in the debate. The actual participants, however, were only a small sample of the population the decision would affect, a tiny elite whose predilections have shaped the world of their successors. Once this is recognized, a commitment to freedom and democracy demands a review of the decisions, not by a scientific coterie but by everyone who has an interest in the outcome—in short, by any or all of us.

My earlier discussion of the entanglement of values and Science provides a way of rebutting this line of thought. Feyerabend and Kuhn were correct to question the power of "method"—or, at least, extant formal versions of "method," alleged theories of rationality—to warrant the major transitions that have occurred in the historical development of the sciences. Those transitions were made by scientists in light of their judgments about relative success in problem solving, judgments embodying probative schemes of values. When they are examined closely, you see how difficult it eventually became for those on the losing side to formulate any adequate scheme of values that would sustain their preferred perspective. *Ideal deliberators reviewing those transitions would be expected to endorse the judgment of the majority in resolving the debate.* That endorsement provides grounds for attributing expertise differentially, for supposing that scientists who have thought hard about the predictive and explanatory successes and failures are in a better position to judge than people who do not know these things. Such attributions of expertise are the expression of genuine democratic values. Feyerabendian "mob rule" is a vulgar surrogate.[1]

To resist inserting vulgar democracy into the context of certification does not settle all the issues about how democratic values might affect what is taken as public knowledge. Fear of the tyranny of ignorance should not blind us to the possibility of a tyranny of unwarranted expertise (against which Feyerabend overreacted). Even if historically oriented philosophical scholarship can show (as I think it can) that, in a small number of important episodes, the certification of ideas of past scientific communities should be endorsed, we should consider whether there are serious possibilities that processes of certification can sometimes go awry. Are there general features of the sciences, as they are practiced, making it likely that, in some particular areas, particular "findings" are wrongly certified, inscribed "on the books" even though an ideal deliberation would determine certification premature or even illegitimate?

It is easy to think there might be, to imagine instances in which a particular ideological agenda or pervasive prejudice inclines members of a scientific subcommunity to favor particular hypotheses, to overrate certain kinds of evidence, or to manufacture evidence. Imagination is not even necessary, for there are well-known examples from the recent past: Sir Cyril Burt's "evidence" for claims about the fixity of intelligence, defenses of the harmlessness of tobacco, and many more (Gould 1981; Oreskes and Conway 2010). The core of the trouble identified in chapter 1 is the suspicion of many people that cases like these pervade many fields of research. Even if they do not, democracy can function well only when it is clear to all that they do not—Science, like Calpurnia, must be above suspicion.

It is useful to divide the potential difficulties into a number of families. Certification might go astray as a result of deliberate dishonesty. Alternatively, potential items of new knowledge might be unwarrantedly accepted (or wrongly rejected) by sincere people who misjudged the evidence. Misjudgments might come about because those people had an inadequate general view about acceptance and rejection of novel proposals: they might discount particular kinds of hypotheses or be biased in favor of some type of hypothesis; they might dismiss some sources of evidence or wrongly rely on others. Or they might have an adequate general view but misapply it in a particular instance. I shall suggest that the possibility of *pervasive* misjudgment is especially pertinent, particularly relevant to the current erosion of scientific authority, and most of this chapter will focus on it. First, however, it is worth considering the threat of deliberate dishonesty.

23. FRAUD AND MISREPRESENTATION IN SCIENTIFIC RESEARCH

Until quite recently, there was widespread confidence that it would be hard to succeed in scientific fraud. Not because scientists were deemed an unusually virtuous bunch, their white coats symbols of inner purity. Rather, in line with the seminal ideas of Robert Merton, commentators on science supposed there are norms of conduct, socially applied and embodied in prevalent practices, that would eliminate the opportunities for would-be cheats. Specifically, the practice of replicating findings would guard against the endorsement of results based on nonexistent or fabricated data. Exposure of a number of examples of fraudulent science, particularly in the biomedical sciences, but by no means confined to them, has shown that this optimism was misplaced (Broad and Wade 1982; Judson 2004).

Announced results based on fraud can insinuate themselves into the accepted corpus of scientific findings in two obvious ways. First, too many experiments simply go unreplicated, some because they appear to be variations on a familiar type; others because the apparatus, skills, or resources needed for them are relatively rare. Second, a scientist's failure to replicate a colleague's work may be taken, quite reasonably, as the consequence of a lack of skill. This is especially true when the researcher who attempts replication is a younger scholar, not yet established. If the original result emerged from a laboratory in which years had been devoted to refining a particular experimental system, the failure of outsiders to reach the same conclusions is plausibly explained by their relative lack of experience. It is a familiar fact, well known to scientific researchers, that experiments are often delicate and that skills have to be developed over a long period of apprenticeship (Kuhn 1962; Collins 1985).

These points may leave intact a weaker version of the original hope that replication would serve to expose fraud. *Unimportant* experiments and the banal findings they are supposed to support might be susceptible to fraud, and the mass of scientific results might contain a few falsehoods that never figure in further developments, as well as some correct claims, originally introduced on dubious grounds. Some prominent instances of scientific fraud instantiate this type. The work of Robert Slutsky, for example, was calculatedly unambitious: in producing the extraordinary spate of papers he hoped would secure his career—at one stage his laboratory issued new research papers at the rate of one every ten days—he deliberately tinkered with well-known experimental systems, varying them to generate predictable results, and then concocting data to suit (Engler et al. 1988). Nobody bothered to replicate Slutsky's experiments, and, even if somebody had taken the pains to do so, the results would almost certainly have been close enough to those Slutsky claimed to avoid arousing any suspicion. If, however, a researcher wants to play for higher stakes, offering an experiment in support of some heterodox view that might point the field in new directions, the work will attract more scrutiny. Here, it seems, replication will be undertaken and any fraud will be detected.

A moment's reflection reveals that this is not necessarily so. If the heterodox proposal is correct, or even a closer approximation to the truth than the received view it challenges, the fraudulent researcher may be undetected and become hailed as a bold pioneer. Someone who made a wild, but lucky, guess and then dreamt up a very complicated experimental setup, might go to press with a piece of experimental fiction, inspiring other workers either to do the hard work envisaged, or, more probably, to discover different ways of testing (and, given the luck of the guess, confirming) the new hypothesis. If this seems too risky a strategy to attract would-be scientific revolutionaries, it is worth appreciating the many gradations between making up the results in the comfort of your armchair and doing the experiments thoroughly and reporting all and only the data you have obtained (Judson 2004). More likely, perhaps, is the imaginative researcher with a genuinely novel idea who, convinced (rightly, as it turns out) of its correctness, decides to ignore some of the less congenial pieces of data supplied in his experiments. Just that profile has occasionally been attributed to some of the most important figures in the development of a field—most notably to the allegedly selective botanist-monk, Gregor Mendel (Fisher 1936).

There is another scenario that appears less likely than the case of the lucky trimmers and fudgers. One way to circumvent the barrier of replication for your novel idea is to make sure that the early replicators will deliver results consonant with those you announce. If you have an ambitious hypothesis, it is likely to attract attention, but if the experiment is sufficiently difficult, or costly, to perform, the zest for replication may fade once a few confirmatory replications are in. Even if your hypothesis is wrong, worse than the orthodoxy it challenges or the rivals whose merits are debated, you may still receive happy endorsements if you can enlist allies to report their confirmation of your claims. Of course, they do not need to carry out the experiments themselves—all of you can calculate the data you would like to receive.[2] What you need, in short, is a cabal.

As I noted, this sort of collusion in fraud appears highly improbable, and, in general, one should make a sober appraisal of conspiracy theories. Nevertheless, it is worth noting that judgments of probability in instances like this depend on identifying the frequency with which occurrences of different kinds are found, and the trouble with the scenario just envisaged is that it would be very hard to detect if it were to be carried out. Assume, however, that scientific research is not corrupted by anything so extreme. It might still contain less flamboyant relatives.

Imagine that you, and a significant number of others whom you know, are distressed about a particular piece of accepted science, regarding it as encouraging people to desert values you take to be important. You and your friends devote a great deal of time to devising experiments that will, you hope, support a rival view you find congenial. From time to time, one of your procedures turns up suggestive evidence; that evidence is not as clear-cut as you might like, and there are some messy data. You present your conclusions by tidying up the results to make a more compelling case. You understand that operating in this way is not generally a good idea, but here you "know" you are right, that the data omitted are misleading—and the social good you are producing surely justifies your procedure. When others come to replicate your experiments, they sometimes question your conclusions, citing the sorts of data you have omitted; since some of the replicators, however, are your friends and allies, devoted to the same cause that moves you, they, too, incline to take just the steps you took—the findings are probably misleading and should be dismissed. In this way, you may not be able to dispose completely of the piece of science you regard as obnoxious, but, with the help of your colleagues, you can muddy the waters sufficiently to create, in the public mind, an impression of genuine debate.

Nobody knows how frequently fraud and dishonesty infect scientific research. Replication alone does not suffice to detect (or deter) fraud. What more might be done?

The obvious response to the kinds of cases that have surfaced in recent years, and to the scenarios just outlined, is to introduce procedures for retaining and scrutinizing data. Findings are to be preserved, raw data are to be submitted in support of the statistical summaries typically given in scientific articles, perhaps spot checks are to be run. Plainly these procedures impose burdens, and it is easy to sympathize with honest scientists who believe most of their colleagues share their probity and who resent the unjust suspicions underlying the program envisaged. In accordance with the perspective of chapter 2, a serious judgment about whether any such program would be worth instituting would require replicating an ideal discussion sympathetic both to the harms caused by dishonesty in research and to the burdens placed on honorable investigators. Until far more is known about the frequency of scientific fraud and about the damage it causes, it would seem difficult to conduct any such deliberation. Nor is it obvious how to design experiments to disclose the efficacy of any program, since so little is known about the rate at which fraud occurs when the intrusive procedures are absent.

How worrying is the possibility of scientific fraud? We ought to consider this question with respect to two different time frames and two types of loss. In the short term, cleverly devised instances of the types of fraud just reviewed can wreak havoc with the careers of young scientists, especially the students of the perpetrator and those colleagues who build on the fraudulent work. In the wake of exposure of examples of fraud, these types of losses are, quite understandably, uppermost in the minds of scientists. The potential long-term damage is more severe. If something is "established" by fraudulent means, it would be possible for it to endure as a presupposition of the field, something built on again and again, never questioned, that would skew subsequent research. Only decades, or centuries, later might the error be exposed, leaving commentators to lament the futility of the labor of so many dedicated investigators whose efforts were undermined because of a fundamental error.

The scenarios reviewed earlier suggest a more relaxed attitude to the long-term concerns. Consider the three main types of villains. There are the unambitious grubbers, those who tinker with an experimental design, calculate plausible results, inflate their lists of publications, and make little, if any, impact on the knowledge actually used in their field. There are also the lucky guessers, people who offer bold new ideas without adequate evidence but whose proposals are subsequently supported. Finally, there are the ideologically motivated naysayers, who may entrench incorrect "results" or block the resolution of a controversy. The first two types are certainly unsavory, and it would be better to thwart them, if there were ways of doing so that did not impose burdens on honest researchers. They do not, however, pose great threats to public knowledge—the grubbers because they set their sights so low, and the guessers because the work they omit to do is eventually contributed by others (people effectively exploited by those whose conclusions they confirm and who, wrongly, receive credit). Even if Mendel trimmed and fudged a bit, he has left no permanent damage.

By contrast, the biased naysayers can undermine the work of the honorable majority. Their success depends, however, on the presence of *two* disruptive tendencies: a willingness to cheat, together with a distorted conception of evidence. Perhaps, then, a focus on values that twist scientific judgment can supply a solution to the serious problems caused by scientific fraud?

A different perspective reinforces the conclusion that the intrusion of illegitimate values is the more basic one. Worries about the legitimacy of scientific certification undermine trust in the system of public knowledge. As survey data show, the public does not think of scientists as typically dishonest—those who resist scientific findings accuse researchers of *ideological bias*. Whether the scientists resisted are evolutionary theorists or defenders of the safety of genetically modified organisms (GMOs), the common charge attributes a distortion of judgment: Darwinians are in the grip of a materialist worldview, champions of Frankenfoods have to serve the masters of agribusiness. To restore public confidence in Science, the problem to be solved is that of locating and eradicating the bias of illegitimate values in a way that convinces outsiders that this has been done. As we shall discover in the sections that follow, this problem is theoretically interesting and practically very difficult.

Given these reflections, one might propose a relatively cavalier approach to scientific fraud. It is worth asking scientists to retain their data and to produce records on demand. Perhaps it is even worth setting up a system of spot checks, in which outside monitors arrive at laboratories and require the production of records (analogous to the random drug testing that occurs in some types of athletic competition). Yet any such policing measures should be balanced against the burdens placed on scientific research. As with efforts to battle crime generally, there are conceivable ways of improving rates of detection that would greatly interfere with the lives of honest people: modern technology supplies a wide variety of intrusive methods of surveillance. With respect to the long-term effects of fraud, the likely scenarios are not so harmful as to require intense efforts to police the lab.

That, however, is to ignore the short-term damage fraud entails. Unwitting association with the perpetrator can ruin a promising career, and it is quite reasonable for scientists to worry about the costs borne by colleagues and students. Once this point has been appreciated, it is easier to see how to approximate an ideal deliberation about policy. *For the costs of intrusive policing and of innocent association are identifiable and are borne by the same people.* Members of individual research communities can reasonably decide, in wide-ranging discussions with one another, conversations that should involve vulnerable young scientists, how they wish to balance the risks of unknowing association with fraud against the burdens of monitoring. Outsiders need to be involved only when there is some credible threat that fraud will generate some distortion of policy, and cases of that sort are probably best handled by insisting on extensive replication of the findings to be applied. For in this context, one can reasonably expect replication to be valuable.

Perhaps this is too optimistic. Yet, as we shall now see, even if the possibility of scientific fraud does not call for increased democratization of Science, the possibility that pervasive misjudgment leads to faulty certification does.

24. WELL-ORDERED CERTIFICATION AND IDEAL TRANSPARENCY

When their inquiries succeed, or at least seem to succeed, investigators submit their results to agencies of the public knowledge system, typically journals and other vehicles of publication. Their submissions are reviewed and, if all goes well, accepted for publication. That is the first, and most obvious, phase of certification. Sometimes the certification does not last: subsequent work leads to qualification or correction of what was once taken as knowledge; occasionally, a report needs to be retracted *in toto*. Provided neither of these things occurs, further certification may follow the initial phase. The original finding can be taken up in the work of others, combined with their results to answer some larger, more significant question. Either as a constituent of that larger contribution to knowledge, or, far less frequently, in its own right, it may be given a place in the central results of a subfield, something important enough to be recapitulated in more extensive surveys and textbooks and transmitted to younger researchers. Most submissions do not proceed to this higher level of certification. Vast amounts of knowledge, products of hours, months, years of hard effort, simply vanish—a visit to any site at which organisms have been collected and deposited will bring home this poignant fact in the form of arrays of drawers that have been unopened for decades. That is not to be regretted, for our knowledge must constantly be reorganized, if it is to fulfill its proper purposes.

This schematic review of familiar features of public knowledge prepares the way for thinking about certification. Individual judgments figure in the process of certification, those of the reviewers, the editors, the colleagues who use the findings, the compilers of the textbooks, and the scattered readers who respond to their decisions about future editions, through determining which sources are widely used. For a submission to be certified in the fullest sense—to be *in* the books—a community of inquirers must count it as true enough and important enough. Value-judgments pervade this process.

Those involved in the process of certification will be able to formulate some standards of adequate evidence. These will figure most explicitly in the reports they write when they ask for more work to be done or when they argue against accepting the submission. They will complain that a sample is not big enough, or that it is not clear how the investigation has guarded against some form of perturbation or contamination, or that an alternative hypothesis has been neglected. Within each scientific subfield, researchers share a common set of standards of this sort, and those canons deemed most widely applicable are often explicitly taught—in courses on "methodology." Yet the sum total of what the subfield could collectively make explicit is hardly enough to generate all the assessments made in certifying potentially new items of knowledge. As I have insisted, no collection of formulable canons covers all the judgments made. Just as the investigator must *judge* that enough has been done to go on to the next stage of the inquiry, so a reviewer must *judge* that the summary of the data provided suffices for the conclusion the author draws.

The system of public knowledge is expected to accord with methodological guidelines that are reliable, in the sense that following them would tend to generate correct conclusions—further to certify submissions in accordance with those reliable guidelines, and, finally, to make judgments that go beyond the guidelines by exercising reliable psychological capacities. Developing capacities of that sort constitutes acquiring "good judgment," something routinely expected to be produced by sufficient immersion in a field or subfield.

The concept of reliability figuring here is a thoroughly pragmatic one. Reliable standards and processes are those giving rise to conclusions that are true *enough*, at a frequency that is high *enough*. The levels are set by the contexts in which the conclusions will be put to future work and by the values of the consequences (§4). Because the future uses of so much of public knowledge are quite unpredictable, it is impossible to specify any precise point values. Unless there are good grounds for supposing the knowledge submitted to bear directly on matters of human welfare, an area of inquiry can allow some leeway to individual assessment: editors can tolerate reviewers who are more easygoing and others who are more strict.

To understand the features of reliability, it is worth looking at a particular type of judgment, one that assigns a value to a magnitude. Here we can give substance to the notion of *being close enough to the truth* by thinking of the distance between the value assigned and the true value.[3] Imagine a subfield of inquiry concerned with a particular class of statements of this form, for which researchers envisage an indeterminate set of contexts of future use. That subfield might well contain judges with a variety of profiles. The extremes are marked out by those who are tolerant along two dimensions and those who are strict along both. *Easygoing* assessors accept the largest distance between the value assigned and the truth and the lowest rate of reliability: it is enough for them if the method used has a probability larger than r of delivering a conclusion within d of the correct value. *Strict* evaluators demand the smallest distance between assigned value and true value and impose the highest rate of reliability: unless the method has a chance greater than r^* of yielding a conclusion within d^* of the true value, they will find it unacceptable ($r^* > r$, $d^* < d$). Everybody in the subfield has an approach to evaluation that falls within the intervals set by these values: the researchers all set some standard of proximity to the truth within $[d^*, d]$ and they require that the chance of the method's delivering a value within their favored distances from the true value be greater than some preferred rate within $[r, r^*]$. There will be statements for which judges who agree on the probabilities disagree in their decisions, and for these statements, the acceptance by the easygoing and the rejection by the strict will *both* be considered reasonable. (I doubt if many real cases of scientific judgment allow for specification of probabilities to allow for any such formal treatment; the point is simply to reveal the structure behind tolerance of alternative evaluations.)

Consider now an extension of the ideal of well-ordered science. The context of certification will be said to be well ordered just in case an ideal deliberation would endorse levels of proximity to the truth and of probability of generating truth so that both the general methodological standards enunciated and the particular judgments extending those explicit standards fall within the range of reasonableness determined by those levels. Intuitively, the ideal deliberators are envisaged recognizing the indeterminateness of future uses of a potential addition to the store of public knowledge. They assess the consequences insofar as they can, under the conditions of mutual engagement: not introducing mistaken claims or failing to incorporate different points of view. Given the uncertainties they recognize, they formulate best-case and worst-case scenarios, adjusting their demands on the parameters that figure in reliability so as to avoid foregoing the best-case scenarios and falling into one of the worst-case scenarios. Their conclusions partition the space of processes of certification into cases that demand rejection, cases that demand acceptance, and cases that could permissibly go either way. The area of science is well ordered with respect to certification just in case its explicit standards and its further judgments accord with the partition so generated.

Sometimes it becomes evident that the practice of some area of inquiry is, or has not been, well ordered in this sense. Famous examples come from attempts to study human behavior, cases in which inquirers leapt to premature conclusions in harmony with prior social views: flawed tests were used to charge ethnic groups with intellectual inferiority or to conclude that people with a particular karyotype have a disposition to commit crimes (Gould 1981). In recent decades, primatology was transformed, as it became evident that previous practices were flawed by an exclusive focus on the activities of male members of primate troops (Haraway 1989; Longino 1990).[4] Cases like these reveal how the accepted methods, carefully transmitted to researchers entering the field, may not live up to the levels of reliability they are taken to enjoy, and that the shortcoming results from presuppositions whose validity is grounded in a pervasive scheme of values. Convinced of particular stereotypes about the "natural," "normal," "proper" behavior of particular groups of people, investigators took various things for granted and shaped their research techniques accordingly. If their presuppositions had been made explicit, and had been exposed to a wider discussion, one engaging thoroughly with the people about whom the researchers drew their quick conclusions, questions of reliability would have emerged far earlier. Hence comes an *epistemic* argument for democratization in processes of certification. Representation of a broader set of perspectives within the scientific community has the potential to expose ways in which the methods used by that community are less reliable than they are supposed, and may thus lead to improvements in certification.

The success of many areas of inquiry to achieve conclusions that are maintained over time offers some reassurance that procedures of certification often go well. That does not mean, however, that those procedures could not be refined by critical reflection on them—that has been the way in many areas of research (think of the recognition of the importance of double-blind studies in medical investigations). Nor does it entail that procedures of certification should not be scrutinized by outsiders. Well-ordered certification consists in serving the main functions of a system of public *knowledge*; to wit, storing up information that is significant and true enough. Certification has been judged by how well its procedures do at promoting those epistemic goals. It can also be viewed from an orthogonal perspective, the perspective of *public* knowledge, by asking how well those procedures accord with the ideas about proper acceptance and rejection current within the larger society. Here a different standard is pertinent. I shall call it *ideal transparency*.

A system of public knowledge is ideally transparent just in case all people, outsiders as well as researchers, can recognize the methods, procedures and judgments used in certification (whether they lead to acceptance or rejection of new submissions) and can accept those methods, procedures, and judgments. Intuitively, the public understands how the knowledge amassed is supported, and everyone's ideas about how knowledge should be grounded accord with the way in which certification is actually done. We imagine individual people with their own ideas about the conditions under which to accept or reject potential additions to their beliefs, and we suppose them to be in harmony with one another and with the experts who pursue specialist lines of inquiry. (Because the notion of reliability has loci of vagueness, the harmony does not require that all people use exactly the same standards—there is room for strict and for easygoing judges; they allow the same latitude, and all their judgments fall within the range it delineates.) Harmony not only obtains but is *recognized* as obtaining. Under these conditions, public knowledge is viewed as a collective undertaking endorsable by all.[5]

Plainly the ideal imposes extremely strong conditions, conditions unattainable in practice. Nobody could hope to recognize all the methods, procedures, and judgments underlying the certification of the whole of public knowledge, nor could we expect that those methods, procedures, and judgments would be acceptable to all. Indeed, if the ideal were seriously realized, the point of having a system of public knowledge would be lost. Public knowledge is important to us precisely because it enables us to take over information from others without working through the evidential details for ourselves. Yet there is a serious point to the ideal. Were you to come to believe that, in some areas of inquiry, the methods, procedures, and judgments used in certifying submissions do *not* accord with the standards you view as appropriate in your own activities of belief acquisition, your confidence in the system, at least with respect to the areas in question, would be undermined. You want it to be the case that you could, in principle, probe any part of the public system, and that, were you to do so, you would disclose processes of certification that conform to your standards. You are not special. Public knowledge is set up for everyone and should therefore satisfy the same condition for all. Hence the strong ideal as I have formulated it.

Let us now consider the two ideals, well-ordered certification and ideal transparency in tandem. If both are satisfied, matters are as good as they could possibly be. Certification in the public knowledge system is set up to promote the goals of that system, all people have personal standards for certification that accord with those used in constructing that system—so everyone is completely clear-headed about certification and does as well as possible in private life—and this state of agreement is recognized. If certification is well ordered but there is a serious lapse from the ideal of transparency, public knowledge does well at storing up significant truths, but its procedures are either opaque to outsiders or are appreciated as being at variance with the standards at least some individuals set. Under these circumstances, there can be suspicion of or resistance to the system of public knowledge (see §31 for discussion of grades of opposition). In consequence, whatever its virtues as a repository of significant truth, public knowledge can no longer play its proper role, either in helping those who feel themselves at odds with it pursue their projects, or in guiding public policy. Scientific authority is eroded.

Suppose, however, that certification within the public system is not well ordered but that the ideal of transparency is relatively well satisfied. The vast majority of people adopt standards for modifying belief that accord with the procedures used in certifying public knowledge. As a result, public confidence in Science is high, and policies based on scientific "findings" are well supported. The only trouble is that the certification procedures used in building up public knowledge accept and reject submissions in unreliable ways, so the statements "on the books" are a mixed bag and the policies flowing from them are not altogether successful. Finally, it may be that both ideals fare very badly. The certification procedures within public knowledge are unreliable, and at variance with the standards of many people. Perhaps there is a vast diversity of standards across the population, none of them particularly conducive to reliable certification. Or perhaps there is a mixture, with the possibility that, if the ideas of many different people, including many outsiders, were synthesized, the reliability of the system of public knowledge would be substantially improved. It is not impossible that humanity has spent most of its existence in this last state—and that, with respect to some potentially important areas of investigation, we remain in it.[6]

The judgment that the first state, in which both ideals are realized, is the best needs defense against the challenge that the homogeneity of standards it envisages is antithetical to important values (diversity is good for advancing knowledge and good for individual self-expression)—a challenge that will occupy us later (chapter 8). I shall argue that we are not in this apparently happy state. Perhaps we are in the second condition, equipped with a system of public knowledge that deploys reliable methods of certification but that deviates, at least in places, from the standards many people adopt. Scientists fear that democratization of Science consists in settling for the third state, in which the reliability of certification is sacrificed to the incorporation of popular standards. Yet this is, I think, to misstate the ideas of radical skeptics (like Feyerabend) who can be read as proposing that we are in the fourth state (doing badly with respect to both ideals), and that some synthesis of a broad range of epistemic perspectives would lead to greater realization of both ideals.[7]

25. THE ROLE(S) OF PHILOSOPHY OF SCIENCE: A BRIEF EXCURSION

At this point, it is worth considering a way in which we might take steps toward improving transparency. Suppose it were possible to elaborate a clear and precise account of evidential support. That account could then be used to demonstrate the extent to which the conclusions of various areas of inquiry were aptly certified. Some group of people—call them *philosophers of science*—might be charged with the task of articulating the theory and then scrutinizing various fields of investigation. After delivering their account, they could then show exactly how major parts of inquiry were well certified, according to the principles enunciated, fulfilling the Leibnizian dream of replacing deliberation by calculation.

The traditional aim of philosophy of science is not that of "understanding Science," for "understanding Science" makes no more sense as a goal than "understanding Nature" (§17). Rather, from its origins in antiquity to the present, philosophical study of the sciences has concentrated on specific issues that concern the goals of inquiry and of the successful pursuit of those goals. Thus many philosophers of science have seen themselves as answering some rather general questions: What is a scientific explanation? What is a scientific theory? When is a hypothesis (*h*) confirmed by evidence (*e*)? Questions like these had a very serious point at a time when it was widely held that all sciences could be viewed as fundamentally alike, and that answering such general questions precisely would aid embryonic areas of inquiry. They also have a point in any context in which it is important to evaluate the success of particular sciences and thus to take up the ideal demonstrative work envisaged in the previous paragraph. Yet, for all the effort expended in trying to answer them, the formulations need amending. The history of three-quarters of a century's work has shown conclusively that different areas of Science are methodologically diverse (Cartwright 1999; Dupre 1993; Wylie 2000), and that most of the interesting challenges and disputes within those areas resist the styles of formalization philosophers have wanted to impose.[8]

The fundamental goals of the philosophy of science remain what they have always been; namely, to use reflection on one or many areas of inquiry to improve scientific practice and to make the standards of good inquiry more evident—goals intimately related to the two ideals of the previous section. Once the diversity of the sciences is fully appreciated, positive effects are most likely to be local, although generality is welcome, when it can be achieved. Potential failures of well-ordered certification provide important occasions for philosophical work. Protracted disputes often signal places where a more general perspective can help, as, for example, in disputes about the inheritance of intelligence, or the status of evolutionary psychology, or the credibility of models of climate change. One main task of the philosophy of science is to identify clearly areas of Science in which the ideal of well-ordered certification is badly served—and, if possible, to advance suggestions for doing better. More general philosophical issues, about explanation, theory, or confirmation, are pertinent insofar, and *only* insofar, as they promote the more local work. It is sometimes appropriate to draw on ideas suggested during past decades, and to craft tools for addressing an issue arising within a particular area of Science. When conceptions of theory or explanation (say) are treated as instruments for specific inquiries into the status of a controversial proposal, we can be released from the temptation to ask what theory and explanation are *really*; it is enough to have a set of possible approaches, a toolkit, that includes something apt for the work at hand (Kitcher 2009).

Epistemic considerations, however, lie on only one of the axes on which Science might fall short. Throughout its long history, philosophy of science has neglected questions about the social embedding of inquiry. Perhaps that was understandable when Science had not yet come to dominate the public knowledge system—that is a nineteenth- and twentieth-century development. Perhaps the failure to undertake social issues at the time when contemporary philosophy of science was formed (in the 1930s) reflected the excitement of promising new philosophical ideas (that of logical reconstruction, for instance). The tangled relations now evident between Science and social decision making (one facet of which is the erosion of authority with which I began) call for philosophical attention to issues on another axis, that marked out here by my ideal of transparency. In recent years, philosophers of science, including prominent feminist philosophers of science, have begun to pay the needed attention (Longino, Wylie, Dupre, Douglas) and have recognized the importance of philosophy of science for human use (Cartwright). I have found myself backing into this territory (Kitcher 2001). The present book is a more systematic attempt to explore it.

26. CHIMERIC EPISTEMOLOGIES AND OPAQUE VALUE-JUDGMENTS

Any attempt to address the erosion of scientific authority must come to terms with our lapses with respect to the ideal of transparency. The previous section bluntly rejected a traditional hope for tackling that problem. We have no choice except to consider these lapses piecemeal, as they arise most urgently for us. I shall consider two types of instances.

Many Americans (as well as people in other parts of the world, increasingly in the Islamic world) resist the conclusions about the history of the universe offered by the public system of knowledge. They think cosmologists err in attributing an age of over ten billion years to the universe; they think geologists mistaken when they explain that various rock strata are billions or hundreds of millions of years old; they think biologists go astray in supposing life appeared on earth four billion years ago, that multicellular life began hundreds of millions of years ago, and that mammals are distant descendants of unicellular organisms. All these "established facts" are wrong. Yet (although the opponents would not put it that way) they have been certified by the public knowledge system. Defenders of the system even claim publicly that these "facts" are particularly well-entrenched parts of what "we know."

Although Darwin is typically taken to be the arch-villain by people who oppose the orthodox picture of the history of our universe (call them "Deniers," for short), only one part of the doctrines they oppose can be credited to the author of the *Origin*. The Deniers are not simply at odds with evolutionary biology but with swaths of other areas of inquiry—consensus views about radioactive decay, parts of standard cosmology, and stratigraphy, for example. They suppose that researchers working in a number of different fields have certified proposed contributions incorrectly. How has this happened? The Deniers typically cannot say, for they have not examined the considerations underlying the acceptance and maintenance of the official story as part of public knowledge. Many of them have heard defenses of the consensus view, in which prominent scientists, not obviously untrustworthy, have declared that the controversial theses are as well-grounded as parts of Science with which the Deniers have no trouble: facts about the shape of the earth, the acceleration due to gravity, and the composition of water. They have more or less detailed ideas about how scientists proceed in the fields whose findings they do not deny. Some of them have done experiments and presented evidence for conclusions they hoped others would accept. Consequently, there are some methods, procedures, and judgments, common in science—and, in less elaborate forms, in everyday life—they do not dispute. Despite assurances that the official view of the history of the universe has been certified by just such methods, procedures, and judgments, they demur. How can they be so confident?

Perhaps some of them have read pamphlets, articles, or books by maverick "scientists" who share their doubts. Most have not done so. They know, in advance of reading that kind of "literature," in advance of reviewing the details of the certification procedures actually followed, that it has all gone awry. *For the official story can only be certified if you dismiss an important source of evidence.* That source is, of course, an infallible scriptural text. The Deniers know how to read that text, and, although they sometimes hear about supposed "rival interpretations" or "alternative ways of reading," they place no stock in such suggestions. For them, the words—God's words—are completely clear and definite.

The Deniers do not dispute the methods, procedures, and judgments typical of the sciences they understand and endorse. Those ways of certifying new ideas are perfectly appropriate in their place. They are, however, only part of an adequate view. Deniers share a different conception of knowledge: standard scientific investigations can reveal many things about the natural world, but, where they conflict with revealed religion, they cannot be trusted; for the texts of the scriptures, read as they should be, offer a higher form of evidence that cannot be overridden by our fallible inquiries. This conception of knowledge has a venerable pedigree, for something very like it has been debated for centuries; Galileo's *Letter to the Grand Duchess Christina* is one of the more prominent attempts to oppose it, but the stakes over which Galileo and his contemporaries fought were far lower than those in the twenty-first century opposition.

The Deniers embrace a chimeric epistemology, one including two methods of certifying that can deliver opposing verdicts about acceptance and rejection. Unless they think scientists in several fields are dishonest, or that they have misapplied their methods and procedures, apparently in systematically similar ways (coincidence or conspiracy?), they have to suppose something simpler: scripture trumps secular procedures, the procedures of refined sciences, and everyday inquiry. I shall now suggest that if this chimeric epistemology were brought into the open and scrutinized, it would be seen as a very uncomfortable position.

Start from a very obvious point about the secular procedures scientists—and everyday folk, Deniers among them—use to certify proposed claims. They work. Whether you are trying to figure out what has gone wrong with the plumbing, why your new plants are not growing, whether a potential drug has severe side effects, who was responsible for the murder, what compounds will be formed from a novel reaction, or what loci are pertinent to a hereditary disease, accepting or rejecting possibilities in light of the procedures favored in the sciences, and by their commonsense relatives, is likely to lead to successful interventions—you act on what you have learned, and the plumbing works, the plants grow, and the patients recover without complications. What can be said to support the reliability of certifying claims by appeal to scripture—and, indeed, of overriding the secular procedures when there is a conflict?

Not only is it hard to supply an answer, but there are plenty of obvious reasons to worry about the envisaged strategy. For there are people who do very similar things, whom the Deniers are committed to regarding as dangerously misguided. They, too, override secular standards when there is a conflict with a sacred text or an oral tradition. The trouble is that their texts and traditions are different—or, disturbingly, they read the right book but deviate from the proper sense. The world is full of people who foolishly override the plain evidence of their faculties (their senses and their powers of reasoning), sometimes quite bizarrely: they think that spirits inhabit particular places, that their ancestors return, that there are occult forces in nature, that there are many gods, and their fantasies lead them to absurd conclusions.

How are these benighted people to be distinguished from those with the correct chimeric epistemology, the Deniers themselves? The Deniers say, of course, that their ancestors had an experience of God, that an infallible message was delivered, and that the message has been preserved intact in the scriptures they revere. Sadly, all the world's benighted folk have versions of this story. They, too, are the beneficiaries of a tradition that began with a revelation to distant ancestors, one that has been carefully transmitted to the present. *By what right can the Deniers, or any of the others for that matter, break the symmetry, claiming that their "authoritative text" is the genuine article, that their chimeric epistemology is the correct one?*

The Deniers' predicament is even worse than I have so far made it appear. During the past two centuries, historical, archaeological, and literary studies have exposed the processes through which religious doctrines evolve. Whether they focus on oral traditions or scriptures, they offer the same message. The marvelous stories that lie at the heart of religions, stories about covenants and bodily resurrection, for example, are constructed long after the alleged facts, modified in accordance with prevailing conditions, interpreted and reinterpreted. Religions prosper for reasons radically unconnected with the truth of their central tenets. The experiences taken as "religious" are inevitably shaped by the categories culturally available.

A wide body of investigations, completely akin to those the Deniers would endorse in trying to understand the development of some nonreligious facet of human life, and which they would willingly accept in the case of all religions except their own, proceed from one part of the chimeric epistemology to subvert any claim that the other part is trustworthy—let alone so trustworthy that it should override everything else. Many of these investigations have been carried out by people who began from the Deniers' own religious convictions, devout theologians who have sometimes lamented that their work undermines faith. The chimeric epistemology can be sustained only if the Deniers resolutely refuse to consider two questions: What is the history that stands behind our religious claims? How likely is it that claims generated from that sort of history would be true?

Perhaps at this point it may be suggested that a basic commitment to standards of certification cannot be based in any way on reasons. It is simply a matter of acceptance. That is not, however, the ways in which the procedures of the sciences have come to be as they are. Our prehistoric ancestors probably tried various ways of trying to evaluate items of potential knowledge; our historical ancestors certainly did. The ones we currently adopt were not the early ones they simply committed themselves to and then stuck with. Instead they are the products of a long process, during which people reshaped their standards in light of what they found to work. Commitment to evidential standards is not a blind leap, taken once and for all, but something that evolves.

Deniers may think differently. "So," a devout Christian, or Muslim, or Jew might declare, "I simply *accept* these claims about past events, these doctrines about what people should do and what they should aspire to be. To ask me to provide reasons—or to play clever games that try to show I have no reasons—is entirely beside the point." That avowal is indifferent to any thought of how the appeal to scripture might be evaluated or modified, and how it has power to override procedures of certification Deniers are happy to apply in other contexts. Attitudes of this sort might be allowed if the use of the chimeric epistemology were a purely private matter. If, however, you are going to use your religious attitudes to run your life, if you are going to let religious doctrine guide you to decisions affecting the lives of others, the willingness to leap to a standard for judgment, to commit yourself in the absence of reasons, deserves *ethical* scrutiny. As William Clifford, the late Victorian mathematician and apologist for science, saw very clearly, we do not usually endorse the behavior of people who act without reason, ardently convinced that things will turn out well. In Clifford's famous example, the shipowner whose wishful thinking leads him to send out an unsound ship is rightly held responsible when the passengers and crew drown. The earnest religious believer who supposes that God has commanded him to kill his son, or that religious doctrine requires him to eliminate the ungodly, or that it is wrong to undertake the operations doctors prescribe to save the lives of children, will subordinate ethical maxims he would otherwise use to guide his conduct to the dictates of faith, faith that is admittedly blind, supported by no defensible reason. We should protest that blind commitment, for, if it is allowed to issue in action, it is profoundly dangerous.

The true character of acting from unreasoned faith is revealed when you look at the actions of those who are moved by a different faith, at militant fanatics who aim to murder people who do not conform to their religion, for example. Many Deniers will naturally think of themselves as different, but, as we have seen, there is no basis for holding that the *religious* doctrines they avow are any more likely to be correct than those of other faiths, even of radical and intolerant versions of other faiths. The blindness with which they commit themselves to acting in accordance with their preferred interpretation of a particular text is no different from that of people who would express a similar enthusiasm for the *Protocols of the Elders of Zion*, or who would regard *Mein Kampf* as divinely inspired.

Democratic decision making faces severe threats from the presence of chimeric epistemologies within a society. For suppose a democratic society consists of two groups, diverging not only in their values but also in their procedures for certification. If there are issues that arise for this society in which each group makes its decision according to what it takes as the facts, and if the differing epistemic standards yield incompatible factual determinations, how will the policy dispute be resolved? Whoever loses will be committed to seeing the outcome as based in a faulty conception of the facts, one rooted in a failure by the victors to respond to what the evidence demands. That might be tolerable if the consequences of the rival policies were not at odds on any grave matter, but it is quite intolerable when human issues of the greatest seriousness are at stake.

Consider, for example, debates about the legitimacy of manufacturing stem cell lineages to use in research on diseases that currently darken the lives of many people. As we shall see in more detail later (chapter 9), people who take the proper certification procedures to be those used in the conduct of Science will see the question in terms of bringing into being small clusters of cells, so that suffering people and their families may benefit; those with the chimeric epistemology favored by the Deniers will regard it as involving the intentional destruction of human life, acts of murder. However the issues are decided, the losing side must regard the result as one in which the most crucial evidential considerations have been ignored. There is no resolution that can compromise between the clashing perspectives, and, for the defeated, the considerations advanced against them must seem a travesty of reason. So they are asked to allow policies to go forth in their name, when they must repudiate both the reasons and the conception of reason on which those policies are grounded. Precisely because of the stress thus placed on a democratic society, it is important to work toward a shared notion of public reason. Without it, as we shall see, trust in public policy (not only in public knowledge) is bound to erode.

The considerations just raised would, I believe, have to figure in any ideal deliberation of the standards for certification of public knowledge (including Science as a central part). Given the approach to values outlined in chapter 2, and the conception of democracy advanced in chapter 3, the appropriate standards for certification, those that contribute to democratic values, would be those reached in conversation under conditions of mutual engagement. In any such conversation, the questions typically sidestepped by proponents of chimeric epistemologies would have to be addressed—Deniers would have to confront the ways in which their own propensities for overriding widely shared certification procedures have emerged. To assert their ungrounded commitment to a particular standard, and to claim that others should abide by policies flowing from it, even when those others repudiate the commitment, would be a dramatic failure of mutual engagement.[9] Ideal deliberation would thus endorse the conclusion that methods of certifying claims as part of public knowledge must be thoroughly and completely secular. Public reason can allow discussants to put forward claims that accord with religious beliefs, but *defense* of those claims must be free of any reliance on the tenets of a religious tradition.

In the next section, we shall address questions about how steps might be taken toward the ideal of fully secular public reason, and address concerns that eliminating religious considerations from the public sphere limits the freedom of the devout. Before proceeding to these important issues, I want to look at a different way in which contemporary societies fail to live up to Ideal Transparency. This occurs when people outside the scientific community cannot recognize the procedures used in certifying some part of public knowledge and, to the extent that they can form ideas about those procedures, find them thoroughly opaque.

As with the example just considered, the starting point is a conclusion many members of the population find uncomfortable: "experts" declare that secondhand smoke is dangerous, or that the average temperature of our planet is on course to increase by 2°C by the end of the century (even if we were to act now). Discomfort arises from the anticipation of policies based on the announced conclusion. Smokers foresee a ban on indulging their habit in public places; commuters fear a steep rise in the price of fuel. At this point, there is an important difference with the case of resistance to the official history of the universe. Resisters do not subscribe to a chimeric epistemology enabling them to override the supposed "new knowledge." They simply do not want the announced conclusion to be true. By itself that would not be enough: they appreciate that "wishing won't make it so." Yet the finding has consequences for them, and they need to be sure proper procedures of certification have been followed. Since they have limited access to the basis on which the conclusion is drawn, they wonder if scientific orthodoxy is imposing on them.

The reaction is not unreasonable. When you are told something that has serious consequences for your future, it is natural to ask your informant, "Are you sure?" The situation is easily exacerbated if there are sources of information that deliver a contrary message. If the Resisters hear from other "scientists" that the alleged consensus on the conclusion is not complete, that the conclusions drawn are premature, that those who draw them are people who overwhelmingly accept specific value-judgments—they were antecedently committed to the desirability of eliminating smoking, and to saving endangered species by lowering carbon emissions—the opacity of the value-judgments actually made in certifying the conclusions will matter. Public *claims* that partial value-judgments, those the Resisters oppose (and might continue to oppose, even under conditions of ideal deliberation), have played a role in generating the "consensus" reasonably reinforce doubt. If there are wealthy groups within a society, interested in blocking the threatened policies, it will often be possible for them to recruit "scientific spokesmen" who can sow the seeds of doubt. The opacity of certification and of the value-judgments playing a role in it thus provides the opportunity for many people reasonably to *believe* that the alleged conclusions have been reached through procedures they would not endorse.

Not only can cases of this sort reinforce one another, as the spokesmen for Resistance on different issues paint a coherent portrait of the value-judgments that move liberal, latte-sipping, tree-hugging, European-vacationing, Birkenstock-wearing scientists to jump to conclusions that impose on groups of people they secretly despise, but the overall attitude to Science and scientists gains support from examples in which chimeric epistemologies are in play. The two types are different. In the first, the standards used—properly used—in certification are not those of a significant fraction of the public; in the second, the standards deployed are those that Resisters would endorse, but the Resisters are placed in circumstances where it is reasonable for them to doubt that this is so. Nevertheless, strategies for cultivating Denial and Resistance share common features. The public becomes used to thinking of Science as dominated by people who do not share their values, and who, in light of the errant value-judgments they make, foreclose debate. The same repellent portrait emerges.

27. SUGGESTIONS FOR IMPROVEMENT

Neither type of lapse allows a completely straightforward remedy, but cases of opaque value-judgments are easier. I shall start with them.

A significant part of the difficulty stems from the allergy to value-judgments and the consequent popularity of the ideal of value-freedom. Once that ideal is current, the processes of certification will be widely assumed to be "objective"—or deficient. A scientist who represents the consensus must thus attempt to describe the processes underlying certification, using restricted vocabulary—and this provides opportunities for the opposition. There are all sorts of possibilities for pointing to new research that might be, but has not yet been, done: "scientists" friendly to tobacco invoke some potentially confounding variable and suggest that it is "unrigorous" to draw any conclusion until its effects have been excluded; climate-change skeptics note that the models used omit some potentially pertinent process and claim it is "unwarranted" to believe the alleged effect until that process has been included. A natural response would be to say: "If the conclusions we propose to draw are correct, there are serious consequences for human welfare; if we were to delay, we should risk considerable suffering; plainly, we cannot consider all possible variables; our judgment is that we have taken into account the important ones." That is exactly the sort of thing responsible scientists feel they cannot say—some of them, no doubt, because they believe in the value-free ideal and would be ashamed to flout it so outrageously, others because, while they recognize the role of value-judgments in their certification procedures, know also that this is not the idiom in which they are supposed to talk.

If the debate is to be fully public—and we shall explore in the next chapter some reasons for departing from a supposed democratic ideal of free public debate—it should proceed as clearly and as openly as possible. Scientists who represent the community consensus are profoundly handicapped by not being able to acknowledge the places at which value-judgments are involved in their decisions, and *by not being able to explain those judgments and their grounds*. They come to be seen as leaping to conclusions, fostering suspicion that the values behind the leaps are political values many people would find unacceptable. Instead of being open about why, under the circumstances, the potential damage caused by secondhand smoke or by belching gas into the atmosphere makes investigating the factors opponents cite unwise, they are forced to appeal, vaguely, to "the evidence." Those appeals leave ample room for their opponents to insinuate, and for the public to believe, that what really drives them is hostility to tobacco or a fondness for polar bears.

Postponing issues about the fruitfulness of public debate of technical questions, I suggest that we can make headway with the problem by extending the role of two channels of communication between Science and the lay public, proposed in the previous chapter (§20). One of these envisaged scientists expanding their efforts at public communication; the other suggested leading small groups of outsiders (with diverse perspectives) "behind the scenes." Both routes should be exploited to lessen the opacity of value-judgments. A precondition of doing so is full and extensive rebuttal of the myth of value-freedom: here lies important work for historians, philosophers, and sociologists of science, but it is work that *must* be done in collaboration with, and not in opposition to, scientists from many different fields. Possibly a first task is to cure the allergy of talking about values as it affects members of the community of researchers, *to bring out the clear and important distinction between judging that the data are good enough, given the more or less precisely envisaged consequences for human welfare, and accepting a conclusion because it will make you money or advance your favorite political cause*. Once it becomes accepted that there is a framework for explaining and discussing values, and that some value-judgments can be convincingly defended if they are presented within this framework, replies to the spokesmen of Resistance can be more effective. This candid explanatory approach, whether adopted by scientists who speak and write for the broader public or elaborated in conversations with the representatives who "go behind the scenes," can be supplemented by attempts to make the value-judgments of the people who lead the Resistance more apparent. Again, historians, philosophers, and sociologists of science can serve the cause by exposing the many instances in which *political* values and short-sighted self-interest play a role in fomenting controversy. Important work of this kind has already been done (Oreskes and Conway 2010). It would be well to have more of it, and to make it widely available.

The practical problem posed by chimeric epistemologies is more difficult because the source of the epistemological stance is so intimately connected with the lives of those who adopt it. How, without undemocratic forms of coercion, could a society that tolerates ungrounded appeals to scriptural texts in its public discussions be transformed into one that acknowledged ideal canons of certification, a fully secular public reason? Whatever the arguments given, however powerful or rationally compelling they may be, it seems enormously unlikely that the deeply religious people who embrace the chimeric epistemology described in the previous section will feel the force of those arguments, so they are moved to acquiesce in restricting standards of public certification to the secular. Indeed, it seems unlikely that they will view the presentation of those arguments as anything other than a threat to their proper freedom.

Religion is, and has been, central to the lives of most people who have ever lived. From what we know of the history of the growth and spread of particular creeds, its pervasiveness is understood in terms of the social purposes it serves, and nobody should expect it to disappear without a struggle under the impact of what proclaims itself—accurately, I believe—as reason. For the benefits religion promises to the faithful are obvious, and obviously important, perhaps most plainly so when people experience deep distress. Secular reason does not seem to provide much consolation at a funeral.

Of course, secularism has its own revered figures, people who met personal tragedies without turning to illusory comforts. Hume faced his painful death stoically, persisting in his skepticism to the end; T. H. Huxley, Darwin's tireless champion, wracked with grief at the death of his four-year-old son, refused Charles Kingsley's proffered hope of a reunion in the hereafter. Perhaps these figures should serve as patterns for us all, admirable examples of intellectual integrity and courage that will not take refuge by turning away from the truth, by supposing, with the supernaturalists, that stories about life after death are literally true.

It is crushingly obvious, however, that those most excited by the secular vision—those who celebrate the honesty of spurning false comfort—are people who can feel themselves part of the process of discovery and disclosure that has shown the reality behind old illusions. Celebrations of the human accomplishment in fathoming nature's secrets are less likely to thrill those who only have a partial understanding of what has been accomplished, and who recognize that they will not contribute, even in the humblest way, to the continued progress of knowledge. How can voices celebrating secularism understand what many other people stand to lose if their arguments are correct?

Overcoming resistance, and creating a fully secular public reason, requires more than the reiteration of the line of argument offered in the previous section, more than *anyone's* favorite line of secular argument. Resistance is hardly unreasonable if what you would be left with is a drab, painful, and impoverished life. For people who are buffeted by the vicissitudes of the economy, or who are victimized by injustice, or who are scorned and vilified by the successful members of their societies, or whose work is tedious and unrewarding; for people for whom material rewards are scanty or for whom the toys of consumer culture pall; for people who can unburden themselves most readily in religious settings and who find in their church a supportive community; and, above all, for people who hope that their lives mean something, that their lives matter, the secular onslaught threatens to demolish almost everything. They need reassurances that there will be replacements for what they so obviously stand to lose.

Writing in the 1920s, thoroughly aware that scientific inquiries into the evolution of religious belief had created a "crisis in religion," America's premier philosopher, John Dewey, argued for a new attitude to religion and the religious. We need, he suggested, outlets for the emotions that underlie religion, and this requires the emancipation of the religious life from the encumbrance of the dogmas of the churches, of their commitment to the literal truth of their favored stories. The task is to cultivate those attitudes that "lend deep and enduring support to the processes of living" (1934, 15). Dewey was, I believe, pointing to a position that needs to be developed and embodied in social life if public reason is to be properly emancipated. Any adequate elaboration must start by appreciating the genuine needs that stand behind religion. "It is the claim of religions that they effect this generic and enduring change in attitude. I should like to turn the statement around and say that whenever this change takes place there is a definitely religious attitude. It is not *a* religion that brings it about, but when it occurs, from whatever cause and by whatever means, there is a religious attitude and function" (1934, 17). At the beginning of the twenty-first century, some affluent democracies, most notably the United States, have not achieved the broadening of the religious life Dewey envisaged. For most Americans, the only occasions that cultivate the attitudes supporting the processes of living are dominated by the doctrines of the traditional religions. The forms of Christianity that have been most successful in recruiting new members place heavy emphasis on the full acceptance of dogma, on literal interpretations of the canonical texts, and thus are most committed to a chimeric epistemology. Scriptural myths pervade many American lives because many Americans can find no replacements for the old ways of supporting emotions and reflections essential to meaningful human existence.

None of this is to deny that religion, as it has been elaborated in the substantive stories of the major traditions, is also capable of doing enormous harm. The history of religions reveals not only the consolations of the afflicted and the legitimate protests of the downtrodden but also the fanatical intolerance that expresses itself in warfare and persecution, that divides families, cities, and nations, that forbids people to express their love as, and with whom, they choose. It is possible to appreciate the ways in which the religions human societies have developed have met genuine human needs, without forgetting that the myths they have elevated as inviolable dogma have often been destructive. Dewey saw our situation clearly—the challenge is to find a way to respond to the human purposes religion serves without embracing the falsehoods, the potentially damaging falsehoods, of traditional religions—secular life needs to be made responsive to our deepest impulses and needs.

Rousseau proposed a precondition for a social contract: the parties must share a conception of the common good. Analogously, a democratic society urgently needs a shared notion of public reason, a common agreement on what kinds of evidential considerations count and on their relative weight. Academic writings on democracy often suppose this notion of public reason must be neutral among all private views, as if the secular standards, the view from Science, were naturally paramount. If, however, the epistemology of evangelical Christianity is committed to the overriding authority of the Bible, evangelical Christians cannot accept Science as the single voice of public reason. For religious reasons to be debarred from public discussion is, for them, for policy to be systematically unreasonable. By the same token, if those reasons are permitted to enter—if religious leaders testify, in the name of their scriptures, before policy-making bodies—secularists (and some religious allies) will see public reason as prey to irrationality and fanaticism. Either way, there are bound to be decisions some citizens will feel it their duty to protest.

This situation has serious consequences. Once people have become accustomed to the bifurcation (or dissolution?) of public reason, they can easily warm to the idea that sources whose deliverances support their values are at least as trustworthy as those that claim a unique title to "objectivity." If that predicament is to be avoided, it will be necessary to address large questions intellectuals often avoid, questions about how to find meaning and worth in a secular life, how to cope with adversity without the consolations (false consolations) religions have traditionally promised. Beyond that, social structures are needed to provide the sense of community, the possibility of joint projects, the space for exploring important questions ("How to live?")—all the things that, at their best, religious institutions have offered. Secularization must be thoroughly humane, more than the confident assertion of the claims of reason. It must recognize the needs for community, for social support, for ways of exploring why human lives matter. The transitions made by some European societies show that the task is not impossible, and much may be learned from studying how those transitions have occurred. If democracy is to be taken seriously, secular inquiry must be committed to such study and to elaborating ways of fulfilling the purposes religion has served—for without a shared conception of public reason, we are truly lost.

Chapter 7

APPLICATIONS AND ACCESS

28. USING PUBLIC KNOWLEDGE

By contrast with the issues just considered, the context of application appears very simple. Suppose the ideals for investigation, submission, and certification have been realized. The agenda for inquiry conforms to the conditions of well-ordered science, research has been pursued in conformity to the constraints issued by an ideal deliberation under mutual engagement, and the resultant submissions have been certified by reliable procedures meeting the ideal of transparency. Now the stage appears to be set for the use of the public knowledge to address the problems that figured in the agenda setting. Some of these will be practical difficulties affecting particular groups of people; others may include the satisfaction of curiosity. Surely what needs to be done is to tackle the practical problems and disseminate the "pure knowledge" to all who sought it.

Often, this will be an ideal way in which to use the public knowledge achieved. Not always, however. One obvious refinement of immediate and automatic application recognizes the temporal gap between the decisions to seek the solution to a problem and the acquisition of adequate information. In the interim, both the overall state of public knowledge and the environment in which the problems originally arose may have changed, so that original expectations are modified. In light of everything now known, the decision to address what was perceived as a problem, or at least to address it in the intended way, may need to be rethought: perhaps the investigation undertaken has revealed many general facets of the situation that reorient earlier ideas about appropriate action, or perhaps environmental changes have altered or resolved the problems. Under well-ordered science, the lines of investigation originally chosen were the product of an ideal discussion in the circumstances *then* prevailing. Change in circumstances may alter the conclusions of any such discussion.

It is obvious what to say in response to this possibility. Well-ordered science should be further extended: applications of public knowledge are well ordered just in case they would be approved by an ideal discussion under conditions of mutual engagement *at the time and in the circumstances* when the knowledge for the application becomes available. Usually, there will be sufficient stability to generate agreement between the two ideal discussions. That should not be assumed, however. As some fans of the autonomy of Science emphasize, we cannot foresee the course of research. Section 19 responded to their concerns about well-ordered science by arguing that planning for an uncertain future is often reasonable. It is equally reasonable to recognize that the unpredictability of inquiry makes it wise to reconsider goals and plans.

So far, this is a relatively straightforward approach to applying public knowledge. More intricate issues arise from two potential sources of complications. First, the information acquired may be insufficient to solve the problem originally posed, while suggesting that the problem is so urgent that immediate action of some form is necessary. Second, there may be serious difficulty in enabling people outside the scientific community, or even those outside a small subfield within the scientific community, to understand the information—complete or partial—acquired. The present chapter is concerned with these complications. As we shall see, they frequently interact with one another.

Start with a taxonomy of potential scenarios, in which investigation, assumed to conform to well-ordered science, produces some partial results.

Application postponed. Inquiry results in a state of public knowledge in which there is reason to think an immediate attempt to use the partial information acquired would be unlikely to succeed (or would involve serious risks), but that information suggests lines of future research offering hope of a more complete solution (one that would have much greater chances of success or would greatly diminish the risks). An ideal discussion under conditions of mutual engagement would conclude that it is best to pursue the envisaged lines of inquiry, rather than trying to act immediately.

Consensus on urgency. Inquiry yields a state of public knowledge in which there is not enough information to offer a reliable solution to a problem, but there is a well-certified conclusion to the effect that inaction would have severe consequences. An ideal discussion under conditions of mutual engagement would conclude that immediate application is needed, even though risks are involved.

Debate about urgency. Inquiry produces a state of public knowledge in which there is debate about the urgency of a problem. Some investigators think the problem is so urgent that immediate application is required; others think it would be better to postpone application and seek more information. Both sides agree that *if* the problem is as urgent as those who press for immediate application believe, something must be done at once.

Lack of access to pure knowledge. Inquiry answers a question, included in the agenda of well-ordered science because it expresses a widespread form of human curiosity, but the answer cannot be understood (or understood as the correct answer) by the overwhelming majority of people.

Lack of access to information relevant to a possibly urgent problem. Inquiry culminates in consensus about urgency or debate about urgency, but crucial parts of the information pertaining to the urgency of the problem or the possibilities of application cannot be understood (or understood as correct) by the overwhelming majority of people.

Lack of access to information about a problem ideally postponed. Inquiry leads to a situation in which application would be ideally postponed, but information pertaining to the need for postponement cannot be understood (or understood as correct) by the overwhelming majority of people, including the overwhelming majority of those most deeply affected by the problem.

It would be wrong to suppose these possibilities exhaust all the scenarios of interest. I introduce them because they arise frequently within contemporary societies, generating problems for the division of epistemic labor and for the crafting of sound policy. Within biomedical research, for example, investigations often initiated with the hope of curing or treating a disease—and advertised to the public as seeking this goal—often follow the scenario of *application postponed.* Frequently, this scenario is coupled to some form of *lack of access*, with the readily understandable consequence that people afflicted with the disease (as well as those who care for them) cannot appreciate the grounds for postponement. Similarly, with respect to climate change, the pertinent scientific community, those granted authority by a proper division of epistemic labor, would take inquiry to have followed *consensus on urgency*. Even if this were appreciated by the citizens of some democratic society, or by the entire human population, the situation would still be problematic because *lack of access to information relevant to a possibly urgent problem* also obtains, with the consequence that the public cannot seriously engage with debates about what immediate steps should be taken. From the perspective of many people, however, inquiry has followed *debate about urgency*. Since the public condition is *lack of access to information relevant to a possibly urgent problem*, there can be no serious engagement with a debate about urgency, let alone about plans for immediate action (given that the problem is indeed urgent).

More generally, the complications of these scenarios, explored in the following sections, bear on a wide variety of occasions in which "dueling scientists"—and dueling politicians who recruit them—argue before a public that has no way of assessing the merits of rival arguments. We shall also consider important ways in which public discontent with Science can arise and the hollowness of the frequently touted idea that Pure Scientific Knowledge is a major public good.

29. THE NEED FOR IMPROVED ACCESS

Previous discussions of how steps might be taken toward realizing the ideal of well-ordered science have suggested local channels through which outsiders might learn more about the results of inquiry (§§20, 27). I now want to argue that lack of access to the results of inquiry causes a number of serious problems—deficiencies in democracy—and that there needs to be a more general program for improving public understanding of the major ideas and achievements of the sciences. At the very end of this chapter, this will be elaborated in concrete proposals.

The heart of the view of democracy developed in chapter 3 is the idea of a citizen who can exercise some control over public policies that would affect his central life project. If there is certified public knowledge bearing on the likelihood that particular strategies would promote, or retard, that project, it will be important for the citizen to be able to have access to that knowledge. When citizens vote in ways that set at serious risk the outcomes they most deeply desire, precisely because they do not understand an item of public knowledge or do not understand the correctness of that item of knowledge, there is a deep failure in the system of public knowledge and a consequent lapse of genuine democracy.

People's concerns for their descendants are likely to be ill-served if they do not appreciate well-grounded scientific information about future conditions on the earth. Their interest in their own health and the health of their children is ill-served if businesses are allowed to use misleading ways of labeling their products: the rise in obesity in the United States, the United Kingdom, and elsewhere surely has many causes, but one contributing cause is ignorance of the exact meaning of the vocabulary used in discussions of nutrition and the clever exploitation of that ignorance by purveyors of processed foods.

A second problem resulting from limited access to public knowledge stems from the misleading advertising issuing from within Science itself. It is understandable that researchers within a particular field should be enthusiastic about the prospects of the investigations they undertake—understandable even that they should incline to their particular versions of the Steinberg-Upper-West-Side-view-of-the-world (§19). Groups of people who suffer from, or are at risk for, particular diseases, compete to persuade the National Institutes of Health to assign research funds to the medical conditions that concern them—and the groups with the deepest pockets typically win.[1] In the absence of the atlas of research possibilities (§20), actual discussions of what lines of inquiry should be pursued involve the clash of competing voices. The pressures to overstate are obvious: "If we build the Superconducting Super Collider, we shall fathom the fundamental secrets of the universe," "If we devote ourselves to nanotechnology, we shall break through into a new generation of machines," "Recent advances in neuroscience enable us to unravel the complexities of the human brain and understand the nature of consciousness," "Mapping and sequencing the human genome will deliver cures and treatments for countless diseases." In each of these instances, there are real benefits, important benefits, that would flow from the research envisaged. Without any serious way of assessing them soberly (lacking an atlas of scientific possibilities), the only available option seemed to be to promise far more than could reasonably be expected.

Most citizens—and politicians—have little more chance of cutting through the hype than they do of recognizing the blandishments of the fast-food labeling. Limited access to scientific knowledge, scientific illiteracy, is problematic in advance because lines of research tend to be chosen on grounds quite remote from what they are likely to deliver. After inquiry has proceeded, after it has been *successful*, its benefits go unappreciated, as a result of the same scientific illiteracy. Mapping and sequencing genomes has taught biologists all kinds of interesting things, providing information that may eventually transform our understanding of development and metabolism and enhance medical practice. As was foreseen, this line of research has provided diagnostic instruments and ways of testing susceptibilities to various types of disease—and, as was also foreseen, the cures and treatments are likely to come slowly. People who suffer from particular diseases—particularly people belonging to families in which a disease is prevalent—and who have generously collaborated with researchers seeking loci that affect that disease are understandably disappointed when the—"successful"—research brings so little relief. Their disappointment can easily harden into resistance to or alienation from Science (§31).

A third consequence of limited access to the system of public knowledge (specifically to the part known as "Science") is its undercutting of a popular strategy for generating enthusiasm about Science. Many of those who write most eloquently about the sciences, and who do excellent work in increasing public understanding of important contemporary results, celebrate Science as a major (the major?) accomplishment of our species. The sciences are taken to provide, independently of any applications, a "pure" understanding of the universe (or, more accurately, of some aspects of it) that is worth having. "We" now understand many things that baffled our ancestors, "we" can contemplate nature without mystification, and this is valuable for "all of us."

But who are "we" who enjoy this happy state? Only a very tiny proportion of the community of scientists can legitimately take themselves to have made any serious contribution to the ability to understand the natural world; although many scientists might be able to appreciate part of the collective understanding, their capacity is not widely shared by outsiders. A sense of active contribution is surely extremely rewarding, a capacity for passive appreciation somewhat less so. Among researchers, this capacity is often limited by specialization, and, for the broader public, even for those who benefit from the efforts to make scientific findings more accessible, it provides only fragmentary and relatively superficial understanding. To the extent that the "pure knowledge" provided by Science is a public good, it is not widely enjoyed.

Improving access to public knowledge in general and to Science in particular would advance democratic ideals in three ways: by giving people greater chances of promoting their projects, by providing them with a more realistic understanding of what they can expect inquiry to provide, and by giving substance to the thought of "pure knowledge" as something valuable for all people. The remainder of this chapter, however, will be concerned with a different benefit, one closely connected with the scenarios of the past section. Flawed access to Science interferes with policies for applying the public knowledge that has been acquired. As we shall discover, this frequently takes the form of substituting a semblance of democracy for the promotion of democratic values.

30. SCIENCE IN PUBLIC DEBATE

Difficulties in application frequently arise when there is, or there is claimed to be, an urgent problem for which existing public knowledge supplies no consensus solution. The focal situations are those exemplified by scenarios of §28: *Consensus on urgency* or *Debate about urgency*, combined (in either case) with *Lack of access to information relevant to a possibly urgent problem*. If there is a *consensus on urgency* among the community of inquirers within whose research field(s) the focal question(s) arises (arise), that consensus may be broadly recognized, or it may be challenged. In the latter instance, the challenge will typically result from disputes about the division of epistemic labor. Scientists outside the field usually viewed as the home of the questions of concern will claim legitimacy for their perspectives on the issues—as when prominent physicists declare their competence to pronounce on climate change. Even when there is a *recognized* consensus that a problem is urgent, difficulties may be caused if there is divergence of opinion about how to respond to it. When there is *debate about urgency*, disagreement typically proceeds at two levels: first as a dispute about whether the problem is urgent, and second, assuming that it is, concerning what to do about it. A similar two-tier structure is present even when there is *consensus on urgency*, if that consensus is not widely recognized.

The general structure of these predicaments is easy to identify. They involve (1) a predicted future state (or a likely future state) assuming inaction; (2) a value-judgment about that state (it is bad and should be avoided); (3) a variety of proposals about what might be done to avoid the future state; (4) a variety of value-judgments about the goodness or badness of the proposed measures. In *consensus on urgency*, there is agreement within the scientific community with respect to (1) and (2); if the consensus is recognized, that agreement extends outside the scientific community; if it is unrecognized, the public (or at least some segments of it) challenges the conjunction of (1) and (2); in *debate about urgency*, disagreement can affect any or all of (1)–(4). (The contemporary discussion of climate change exemplifies this last scenario.)

Consider two different procedures for coping with predicaments of this general type. The first would erect a barrier between the scientific community and the public until the internal differences had been hammered out. Scientists are to work out what they take to be the best practical approach: perhaps deciding that the problem is not urgent and postponing application of public knowledge; perhaps counting the problem as urgent and reaching a single proposal for dealing with it. If their discussions do not end in agreement, the second best will be for them to find something with which all can live or, if that fails, to proceed by majority vote. The second strategy—the "democratic" approach—opposes any barrier behind which the scientific community hides its internal differences, opting instead for debate in a fully public forum.

I shall argue that, in situations marked by lack of access to crucial information, *neither* of these strategies can be considered as satisfactory. There are potential circumstances under which one or the other might satisfy democratic ideals, although it is very hard to see how those circumstances might obtain. The "democratic" second strategy would be appropriate only if the epistemic defects were made up; that is, if all the participants in the public debate fully understood the information available and were also qualified to make the kinds of judgments about likely outcomes reached by the experts. Even though we can *improve* access to scientific information, the improvements seem very unlikely to go that far. The apparently elitist strategy, in which scientists work out their difficulties "behind the scenes," would also be adequate if (and only if) the community of discussants contained representatives of all human perspectives—something an actual scientific community is never likely to manage.

The reasons for rejecting the first strategy run parallel to those urged in resisting autonomy in decisions about what lines of inquiry to pursue (§19). Value-judgments are crucial to crafting a policy about what to do—note the presence of (2) and (4)—and those judgments should embody the perspectives of all who would be affected. By the same token, the second, "democratic," strategy has just the disadvantages attending vulgar democracy; to wit, the tyranny of ignorance. Not only is this likely to arise with respect to the factual claims debated (for example, claims about the probability that a particular consequence will flow) but also in the expression of raw (untutored) preferences.

Attention to the elements of these predicaments will make these points more obvious. Rival scientists envisage different scenarios for the future. The probabilities ascribed to those scenarios are often different, and there is disagreement about the human consequences flowing from them. (Think, for example, about the many different scenarios currently advanced about climate and weather changes in future decades.) Estimates about the chances of the various outcomes, where they can even be made, depend on alternative ideas about the most important causes. Outsiders have little hope of assessing the merits of the clashing claims. Although one party to the dispute might make a better case than its rival—that is, given the information available, its judgment is more likely to be true—no knockdown argument can be offered, and opposition is not unreasonable. The considerations favoring the somewhat superior perspective are technical and subtle, not accessible except to the specialists.

Reflection on the historical episodes that once inspired skepticism about rationality in the growth of knowledge is useful here. Revolutions need time for their resolution (consider the decades of debate about heliocentrism and Lavoisier's "new chemistry"), and, along the way, the competing protagonists can point to a mix of successes and failures. At the end, on the account I proposed, the balance of achievements and unsolved difficulties is so skewed that it becomes impossible for the losing side to propose a probative scheme of values to support its preferred approach. At earlier stages, though, even though a fully informed observer, uncommitted to either perspective, would judge one of the sides to be ahead—"on points," as it were—the decisive blow has not been delivered. If there were some external constraint, demanding a verdict to be issued at an intermediate phase, the best that could be hoped for would be the evaluation of this fully informed observer.

Throwing the debate into a public arena is highly unlikely to replicate the considerations that would move the observer. If the observer's final judgment were achieved, that would be a matter of brute luck. Theoretically, the public's lack of access to crucial information might lead you to expect no correlation (or anticorrelation) whatsoever. In practice, the choice of the public is likely to depend on the rhetorical skills of the participants. I shall suggest shortly that this might make the eventual verdict inferior to one obtained by blind guessing.

Talk of a *single* "impartial" observer oversimplifies the predicament, for, as noted, it is permeated by value-judgments. Some of these are very obviously concerned with the extent to which various outcomes are counted as good or as bad, but others are embedded in the assessment of risks. The debating scientists differ on what causal factors are important, regard different samples as adequate and inadequate, are variously averse to risks of diverse kinds, and hence disagree about which factual claims are well enough supported. We would do better to replace our ideal observer with a larger group, one whose members shared the observer's lack of investment in "winning the fight" but who represented different human perspectives. Once again, the appropriate ideal expands the framework of well-ordered science, envisaging an ideal conversation among mutually engaged parties who know in advance that no decisive considerations can force judgment (there is going to be no knockdown argument) but who look for a way of proceeding, in a mixed evidential situation, that can be acceptable to all.

The most obvious way to nudge our practice in the direction of this ideal would be to expand the role of the groups of citizen representatives envisaged in §20, the small groups who are led "behind the scenes" and tutored sufficiently to offer ideas about which lines of investigation would be worth pursuing. Here they are assigned a more difficult task, one that involves resolving a dispute among reasonable—although possibly not *equally* reasonable—parties. In accordance with the ideas of chapters 2–3, these groups should be as broadly representative as is compatible with their discharging the role assigned to them. Effectively, this proposal amends the *first*—elitist—strategy, for it erects a barrier between the research community and the public, a barrier behind which decisions are to be made. The difference is that a select few are to be allowed to cross the barrier, and, after tutoring, to serve as representatives of the broader constituency.

31. THE SHIBBOLETH OF "FREE DISCUSSION"

To commend this approach to those difficult choices, when, despite our awareness that our information is partial, we have to consider the possibility of immediate action, may seem insufficiently democratic. One of the great themes in discussions of democracy is the importance of free and open debate. The idea that there should be no limits on freedom of discussion receives passionate defenses from some of the most thoughtful and eloquent writers in our language. Here, for example, is Milton: "And though all the winds of doctrine were let loose to play upon the earth, so Truth be in the field, we do injuriously by licensing and prohibiting to misdoubt her strength. Let her and Falsehood grapple: who ever knew Truth put to the worse, in a free and open encounter" (1963, 318–19).

Or, two centuries later, Mill's pithy formulation: "[T]he peculiar evil of silencing the expression of an opinion is, that it is robbing the human race; posterity as well as the existing generation; those who dissent from the opinion, still more than those who hold it. If the opinion is right, they are deprived of the opportunity of exchanging error for truth: if wrong, they lose, what is almost as great a benefit, the clearer perception and livelier impression of truth, produced by its collision with error" (1998, 21). Familiar as these passages are, and happily as we nod our assent in reading, it is an interesting fact that each of them elides a different point, while both of them share a common assumption.

Milton makes explicit a presupposition Mill takes for granted. The encounter, in which Truth and Falsehood grapple, must be fair and open. Replacing the metaphor with more literal language: public discussion of controversial issues must occur in such a fashion that those who make the final assessment of which, if any, of the rival opinions is correct must be able to do so on the basis of the evidential support accruing to each. Call this the condition of *evidential harmony.*

Mill likewise is open about something Milton fails to specify. He supposes that those who benefit from the freedom of discussion are human beings—possibly considered as a collective (as the talk of "the human race" suggests), perhaps viewed as individuals (the particular people who obtain the "clearer perception and livelier impression of truth"). There is no suggestion that different people or groups of people might be affected differently, either by the course of a public discussion or by its outcome. Mill supports the condition of *equal benefit.*

Neither Mill nor Milton says very much in the quoted passages about the nature of the benefit that might come from arriving at the truth. The essays from which the quotations are taken indicate that neither of them is concerned with truth as an ultimate value. Mill's treatment of freedom of discussion is preceded by a chapter in which he is much concerned to defend what he takes to be a more fundamental freedom. "The only freedom that deserves the name, is that of pursuing our own good in our own way, so long as we do not attempt to deprive others of theirs, or impede their efforts to attain it" (1998, 17). Mill's conception is in line with the approaches to values and to democracy adopted in chapters 2 and 3 (hardly surprising, given the debt of those discussions to his ideas!): gaining knowledge is important for the realization of this freedom, and freedom of discussion is important for the role it plays in enabling people to gain relevant forms of knowledge. So there is a hierarchy of values, in which a form of self-determination occupies the most fundamental level, in which knowledge obtains its value from the promotion of the most fundamental value, and in which free discussion is seen as valuable because it leads to knowledge.

Milton also accepts something like the hierarchical view, although he develops it differently. Milton's view of the fundamental freedom is the freedom to discover the path to God. Unsurprisingly, given the social environment in which he grew up (§15) and the occupations he pursued, he accepts a hierarchy of values with a Christian framework superimposed. Mill revives the Greek question "How to live?" and subtracts the Christian framework.

The "democratic ideal" of free and open discussion should not be taken over uncritically, without considering the preconditions appreciated by its most thoughtful and passionate champions. Given the approach adopted in chapters 2 and 3, or given the kindred hierarchical views favored by Milton and Mill, free and open discussion is valuable *when the discussants meet the condition of evidential harmony*. When the questions needing discussion are technical, and when most people suffer lack of access to crucial information, evidential harmony fails. Many contemporary citizens live in societies in which there is massive ignorance about all sorts of things that affect those citizens' projects. Indeed, public ignorance comes in grades, with plenty of people at the most extreme grades.

Consider any policy decision that affects the project an individual aims to pursue and the issues pertinent to that decision. The very best position that person could be in would reside in an ability to recognize the questions that are relevant and a state of sufficient knowledge to resolve them. Actual people lapse from this optimal state along one of two dimensions. The more basic problem stems from not appreciating what needs to be settled in order to formulate a reasonable policy: those who do not see what sort of information is needed are truly in the dark. Yet even people who understand the relevant questions may not know the answers. People in this predicament suffer various grades of ignorance.

The mildest grades are those of remediable ignorance. You are in a state of directly remediable ignorance if you have enough background—acquaintance with the concepts and principles of the pertinent branch of knowledge—to be able to consult sources and learn for yourself (or work out for yourself) the answers to the pertinent questions. Your ignorance is indirectly remediable if you lack the background that would enable you to hunt down the answers for yourselves but you know enough to identify reliable people—experts—to whom you can turn for enlightenment: you have an adequate conception of the division of epistemic labor. Much of our thinking about the value of free discussion presupposes an individualistic epistemology that takes citizens to suffer from directly remediable ignorance—their natural wit enables them to see Truth as the winner in the encounter. In the contemporary world, that presupposition is so far from reality that it is hard to credit it even as an idealization. Yet even when that fact is acknowledged, reflections about public discussion take only a tiny step toward incorporating the social character of most of our knowledge, assuming that, when ignorance is not directly remediable, it is indirectly remediable—and, in consequence, continuing to make a shibboleth of "free discussion."

Irremediable ignorance abounds. Many citizens understand that they do not know enough to address technical questions themselves (even if they do some reading), and are also quite confused about who, if anyone, might have expertise. Even thoughtful people can easily be seduced into thinking there are two sides to questions about the history of life, competing "experts" who make claims a lay audience has no way of sorting out. In debates like this, skillful rhetoricians will tap into attractive themes of populism and tolerance. The confusions occurring in this debate are further reinforced by widespread tendencies to favor a chimeric epistemology.

The simplest case of irremediable ignorance—signaled by the confession that someone does not know whom to believe—can deepen into more problematic conditions. First comes resistance, the thought that the alleged experts, those who flash the supposedly brightest credentials, are motivated by considerations quite antithetical to the projects of the folk. "Expertise" in a particular area—say, the history of life—is resisted because it rests on a secret ideology, an atheist agenda (plainly this connects with the problem of opaque value-judgments [§26]). The extreme grade of resistance is alienation, a condition in which large portions of what the "elite" takes to be well-established are rejected on the grounds that supposed "expertise" is just a front for undermining the knowledge (wisdom?) of the "people," and for promoting goals that are opposed to genuine democracy.

These grades of ignorance can readily be recognized in American debates about the history of life, but they are by no means confined to a single country. In other parts of the world, they are most visible in controversies about genetically modified organisms. The latter disputes are considerably more significant—especially for the lives of the world's poor—than are worries about what biology teachers should tell schoolchildren, but both conflicts pale in comparison with the discussions over global warming. Misunderstandings about evolution are hardly one of the world's great tragedies. Failures to accept the reality and the foreseeable consequences of anthropogenic climate change bear somewhat more directly on the life projects of many people.

As I have frequently remarked, the well-being of children and grandchildren is central to the variant conceptions of the good life shared by almost all parents. We might thus conclude that accurate information about the climatic conditions in coming decades, and about the resultant possibilities for agriculture and defense against the elements, might be profoundly relevant to political decisions made now—that voters for whom the lives of people who will live through the 2050s are matters of central concern would want to choose candidates who propose to take steps to forestall huge future problems. Just as those voters would express their political freedom by supporting policies that will limit the dangerous deterioration of the environments their descendants will inhabit, so, too, their freedom—the only freedom deserving the name (Mill)—is undermined if the condition of evidential harmony is not met. Striking instances of the problem of unidentifiable oppression occur when elite groups construct a "free discussion" so that information crucial to the life projects of citizens comes to be perceived as unreliable—in effect, the citizens are used as instruments for supporting politicians and policies contrary to their own central interests. When the conditions of public discussion with respect to important and central issues allow for, even foster, widespread states of irremediable ignorance (as well as resistance and alienation), asserting the value of "free exchange of ideas" is no more than the expression of a shallow, habitual piety.

There is little doubt that we currently live in circumstances in which democratic discussion is hampered by prevalent irremediable ignorance on several issues central to the life projects, and thus to the basic freedom, of most citizens. Nor is it hard to trace causal factors playing a large role in creating and sustaining this unfortunate situation. When the channels through which information is distributed to the voting public systematically distort findings that have achieved full international consensus—when, for example, a majority of the newspaper reports on global warming suggest that this is a disputed question—irremediable ignorance is bound to flourish, and Falsehood has built-in advantages in its grappling with Truth.

Behind these exercises in obfuscation, it is easy to recognize a disastrous flaw in the public institution for disseminating information (Herman and Chomsky 1988; Leuschner 2011). If one supposes that free markets work magic, irrespective of background conditions, one may be inclined to think that media will become self-policing, that no ethical constraints are necessary. Television stations, websites, newspapers, and magazines that transmit falsehoods will wither under consumer selection. Disinformation thrives, however, because there is no invisible hand that will favor the responsible and the accurate. A "news source" can gain adherents because it is entertaining, because it tells fascinating stories about subjects that have no bearing on central issues, because the falsehoods it spreads are hard for members of the public to identify as such—and hard in part because of the large number of channels through which falsehoods are distributed—because its package of information proves consoling to people with a specific prior political view, because it offers slick critiques of its rivals, and so on and so forth. Economists recognize the problems of markets when there are asymmetries in information (Akerlof 1984). If the "goods" brought to market are supposed to provide information, and if they frequently mislead, then the ability of any market mechanism to keep public opinion on track is likely to show progressive decline. Irremediable ignorance will spread, and, as people become more and more misguided, they are in an ever worse position to appraise the messages they receive.

Hence, there are conditions, apparently present in contemporary societies, under which public discussion would be unlikely to enhance the freedom of citizens. This line of reasoning can be taken one step further. Imagine an atmospheric scientist who discovers something that lessens the credibility of a model used by a significant number of her contemporaries to estimate the rate at which sea levels are expected to rise. The scientist knows that there are other models—more complicated mathematically—that support similar conclusions about the potential disappearance of coastlines. She is also confident that, given further research, she and her team could refine the simple model whose estimates have been challenged. That research would take at least a year to carry out. She decides to postpone publishing the finding until she has an amended version of the model.

In her laboratory is an ambitious postdoc. After the lab meeting at which our atmospheric scientist announces her finding and her research plan, the postdoc decides to leak the new result, as well as a report of the discussion, to a news source. The consequences are as we might expect. Within a few hours, a number of media presentations inform the public that a major argument for anthropogenic global warming has been refuted, that the scientific community has attempted a cover-up, and that the situation has only been saved thanks to the courage and integrity of a vulnerable young whistleblower. All this is dressed with ritual phrases commending the virtues of free and open discussion in a democratic society—although only the most high-quality media borrow from Mill and Milton.

Have any ethical mistakes been made here—and, if so, which and by whom? The atmospheric scientist was not wrong to withhold the information from the public; she wisely foresaw the danger that it would be deployed in misleading ways and attempted to do her bit for the promotion of public freedom (of the "only freedom worthy of the name"). If she went astray, her lapse rested on a misjudgment: perhaps she should have had a clearer view of the character of her postdoc and thus been more cautious in disseminating the finding within the lab. The postdoc's decision to publicize the result, however, is at best a naive misunderstanding of the potential damage—perhaps based on too simple a reading of Milton and Mill?—and more probably a self-promoting act that sets his own ambitions ahead of the public good.

Despite its apparent elitism—only a few citizens are taken "behind the scenes" and tutored; only they make the decisions—the extension of well-ordered science to cope with the difficult predicaments we have been considering embodies a deeper commitment to democracy than the proposal to throw the debate into a public arena. Nevertheless, the ideal of free discussion, eloquently championed by Milton and Mill, is attractive and important. It would be better for us to live in a world in which free inquiry and free discussion could be practiced without any threat to more fundamental forms of freedom. Since we do not live in a world like that, the important question is how we attain the ideal state, or at least some closer approximation to it. How can the lack of access to crucial information be overcome? How can the problematic conditions of irremediable ignorance be remedied?

One source of trouble, already noted, is the intricacy of decisions when many different types of factors—"incommensurable" factors—have to be balanced against one another. As studies in the history and sociology of science, from the pioneering work of Kuhn on, have revealed, the considerations underlying many scientific controversies are subtle and delicate: even for communities of experts, resolving debate often takes a long time. When these complex debates are presented in the public sphere, when they are reported by media sources whose profits depend on their satisfying the tastes of people who have antecedent prejudices and no understanding of the technical considerations involved, the notion of "expertise" comes to look increasingly dubious.

Division of labor is central to contemporary societies, and the division of epistemic labor is fundamental to democracy. *As things stand, "free and open public discussion," far from being the expression of democratic values is actually subversive, for it tends to undermine a previously well-functioning division of epistemic labor.* How is that situation to be remedied? The grades of ignorance are signs of a failure of trust—we lack institutions on which people can rely for facts that matter to their decisions. Trust cannot be restored by untrammeled public discussion, for once trust in expertise begins to break down, "free expression of ideas" often erodes further the credibility of those who know. The irremediably ignorant, especially those who are resistant to or alienated from the systems of public knowledge, view the supposedly authoritative consensus of experts as the expression of the ruling ideology. Even if political authorities were to declare that particular people, or groups, are the real experts on various matters, that declaration would only reinforce resistance and alienation.

The best solution to the problems of democracy is more democracy. Where irremediable ignorance abounds, there is nothing to be done but to try to correct it by setting up channels through which some representative body of citizens—including some of those who are most alienated from current institutions of public knowledge—can be brought to understand the consensus achieved by experts, can serve as a representative group for restoring trust in expertise, and can play the role of arbitrator in those difficult predicaments in which experts differ. The presence of that group might eventually restore conditions under which a broader public debate might promote democratic ideals.[2]

I began by considering two strategies for approaching difficult predicaments involving urgency (or proposed urgency) and partial information: an apparently elitist strategy (the expert scientists work out what the best option is) and an apparently democratic strategy (the issue is thrown into a public forum). Both have been rejected in favor of a modification of the former strategy (the "elitist" method) that further extends the framework of well-ordered science. Because that proposal provokes obvious worries about the "suppression of free discussion," I have tried to defend it as an appropriate approach to the problem. I now want to close this (lengthy!) discussion by looking at a potential alternative: might we not amend the second strategy by responding to its crucial defect; to wit, the lack of access to information that transforms "free and open debate" into a tyranny of the ignorant?

Ultimately we might hope to do that, but the route will probably have to go through the more limited involvement of citizens, envisaged in the modified first strategy. Evidential harmony could, in principle, be restored by a massive educational program that would patiently tutor all citizens in the reasoning underlying the positions pitted against one another in technical debates. The potential urgency of some questions would render any such program too slow. Instead, given the pervasiveness of irremediable ignorance and its more pathological forms—resistance and alienation—any solution on an appropriate timescale will require smaller oversight groups, each representing the range of public opinion on disputed issues, which would work closely with communities of experts to understand the state of the disputes and the grounds to which contending parties appeal. If formed in sufficient numbers, these bodies might eventually re-create the trust in public knowledge we have lost, identifying communities of reliable experts and simultaneously setting standards for reliable media reporting.

This last possibility will serve as a final point for this section. Public misinformation—and public suspicion of expertise—abounds in significant part because the channels through which information is currently distributed are responsive to considerations having little to do with reliability. Newspapers, radio programs, TV stations, and websites seek economic profit, and current economic conditions favor creating a niche in which you tell a particular segment of the public the things it most wants to hear. When ignorance is prevalent, media sources may well be most profitable when they serve as reinforcers and amplifiers of ignorance. How can this dismal situation be improved? Not simply by instituting a responsible source of public information, one invulnerable to the whims of the market. Declaring that you have re-created the (old) BBC or that you have resurrected Walter Cronkite would be no more effective than a political edict conferring expertise on particular groups. Alienated people would, quite reasonably, worry that the new institution was a propaganda machine, the organs of some detested establishment. An independent source of information has to earn its credentials. The citizen groups that figure in my extensions of well-ordered science, scrutinizing certification procedures and adjudicating urgent debates, could extend their activity to supervising and appraising sources of technical information. To the extent that they could retain public trust, they could confer trust in independent channels of transmission, curing irremediable ignorance and restoring confidence in a reliable division of epistemic labor. At that point, "free and open discussion" promoting democratic values might become a realistic possibility.

Inspiring writers from the past—Milton and Mill, for example—should not be viewed as supplying directives that can easily be applied to our current condition but rather as offering ideals toward which we might work. Although I have offered some tentative suggestions, it is not obvious how to create the conditions of a free and open encounter in which Truth might grapple with Falsehood, but we surely need a better approximation to those conditions if citizens are to enjoy the only form of freedom worthy of the name.

32. EDUCATING CITIZENS

If, as I have suggested, there are serious shortcomings in most people's access to information that bears on their choices and on their central projects, and if these shortcomings interfere, possibly disastrously, with the functioning of democratic societies, then an obvious possibility is to rethink education so future citizens may be better prepared for the choices they have to make. Education need not be confined to what occurs in schools and colleges: in fact, it had better not be so confined, for citizens will spend most of their lives beyond the classroom walls, and they should, ideally, keep abreast of further discoveries. Democratic societies require a variety of channels from which information can flow from the research community into public consciousness. How those channels are best structured will depend on the capacities for receiving new information that have been set up among citizens. Transmission to those who have left school should be attuned to the skills inculcated in earlier education and to the limitations and lacunae that education has left.

One healthy trend in past decades has been the increasing number of research scientists who have undertaken the job of communicating central ideas in their chosen field to people with no professional scientific training. For many areas of inquiry, including some of daunting technicality, lucid and elegant popular expositions have been provided. The sneers directed toward talented authors and adroit television presenters are far less frequent than they used to be. Yet the boundaries of the audience are evidently limited: the resistant and the alienated are not included. Ideally, the trend to spend part of the time in communicating scientific ideas should continue, involving as many members of the research community who have a talent for it, and it should pay special attention to enlarging the segment of the public to which information can flow. That, of course, involves overcoming the problems noted in the last section.

One very obvious way in which public understanding of the sciences might be improved is through raising the standards of reporting in those newspapers and magazines that, rightly, see the need to inform the public about what is new. In the United States, newsprint journalism is severely handicapped by adherence to two ideas about the proper form of a "popular" article. The first is that the reading public is fascinated by controversy. The second is that human interest is necessary if scientific ideas are to be made palatable. So, for example, the *New York Times*, the nation's premier newspaper, offers each Tuesday an eight-page section devoted to developments in Science and Medicine. The overwhelming majority of its articles conform to a simple pattern: an exciting result or idea has been proposed—material about the proposer—others disagree—material about them—back-and-forth exchanges—further research is needed. Only rarely does the section provide seriously informative articles on established results, information useful to the public (a marvelous exception is an issue completely devoted to genomics and the multiple roles of RNAs). The *Times* is rightly viewed as an outstanding newspaper, and it stands to its credit that it takes the dissemination of scientific research seriously; nevertheless, the articles of its Science section rarely provide enlightenment about issues of public importance, a fact most plausibly explained by the market pressures on contemporary journalism.

As the work of writers like Richard Dawkins, Jonathan Weiner, Brian Greene, and Olivia Judson reveals, it is possible to offer lucid expositions of important ideas and results, accessible to a wide audience, that make fascinating reading. Science journalism does not have to be dull or to woo its readers with tales of "how researchers live outside the lab" and of dueling scientists. The channels of transmission leading from the research community to the general public will not fulfill their important educative function until the standards of Science reporting require journalists to emulate the achievements of the scientists who best communicate the central ideas of their research fields.

The work of transmission would be eased were aspiring communicators able to presuppose a higher level of scientific literacy. That might be achieved if affluent societies with developed systems of public education considered seriously the goals of teaching young people classes that go under the names of the principal branches of Science.

What are those goals, and what should they be? One very important end is to provide some young people the opportunity to go on to careers in scientific research, or to engage in types of work requiring depth of knowledge in some area of scientific inquiry. Commitment to the Millian ideal of finding out what kind of life you want to pursue requires a range of early experiences. Since, in the modern world, a significant number of attractive possibilities require technical scientific knowledge—engagement in research is a principal constituent of many life projects, for example—it would be wrong to deprive young students of the opportunity to follow this route. Whether traditional ways of teaching the sciences embody the best ways of advancing toward this goal, it is clear that they provide experiences enabling students to recognize what would be required of them if they were to aim for a scientific career (either one in research or one that presupposed knowledge of large amounts of scientific material—as, for example, in medicine or engineering). Without the standard elements of scientific education—memorization of vocabulary, problem solving, experimental work, data analysis and presentation—young people would be ill-prepared for a serious scientific career and ill-equipped to decide if this was a good option for their lives. So, at some stage, students require immersion in the details of one science—or preferably a few—the sort of immersion that would prepare them, if they chose, for higher scientific education (and ultimately for a career).

It is well known, however, that most of those who attend classes in the sciences will not follow this route. Once they have discovered that Science is not for them, they often have to suffer through "more of the same," semesters and years of rote memorization, testing on problems, experiments, and all the rest. Many resent these exercises as tedious, and whatever information is acquired quickly slips away. The list of amino acids becomes a blur, any facility at working out the trajectory of a particle disappears; maybe a few skills in titration carry over to the garden or the kitchen. Worse, the lingering impression of Science is of something alien and boring, something best left to others. If people who have been through the post-Science blues are recaptured by writers like Dawkins and Greene, that is testimony to the authors' expository talents.

Giving everyone the opportunity to figure out if he or she wants to aim at a scientific career is entirely compatible with devising a form of Science education that is purposeful and rewarding for the many who come to believe that their lives should go in a different direction. Instead of compelling them to keep tackling problems they find irrelevant or baffling or both; instead of taxing their memories with terminology they will quickly forget; instead of forcing them, year after year, through a weekly regime of experiments; we might encourage them to become happy *consumers* of scientific information. That would equip them for a role as citizens, preparing a far wider pool of readers for the best scientific communicators, an audience for more informative scientific journalism, and an electorate whose choices would not be so deeply affected by lack of access to crucial information.

What do citizens need to learn? Discussions of Science education often emphasize the importance of "understanding the scientific method." Since there is no such thing as "the scientific method," this is a misguided goal, but there are important targets in the vicinity. Students should become acquainted with the phenomenology of scientific practice: they should have experience of what it is like to start work on a question, to design an experiment, to try to make it work, to collect and analyze data. They should also become aware of the types of inferences that need to be made in arriving at scientific conclusions and in resolving scientific debates. Historical studies of past episodes and controversies can be extremely valuable in some of these respects: How could you come to decide that the earth moves? How could you use an inclined plane to find out something about the motion of a falling body, without using a clock? How might you show that combustion is a process in which something is absorbed?

A second important part of Science education for the citizen is to convey the sheer ingenuity and beauty of some of the greatest scientific insights. The authors who are most successful in writing for the broader public provide models for doing this—and there is no reason to withhold the fun from the young. Students whose school years included a course in "Great Scientific Ideas" might carry into adulthood a motivation to learn more, to keep up with the similar accomplishments made in their own times.

These two components of Science education are important parts of achieving an important goal; namely, producing a scientifically literate citizenry. To be scientifically literate is partly a matter of understanding the practice of the sciences enough to recognize that the two sides engaged in today's urgent debate are making intricate and informed judgments, that the matters on which they disagree are delicate, and that the best assessment of what is to be done is likely to come from combining the views of their (expert) community with the perspectives of a diverse group of outsiders. It is partly a matter of appreciating the brilliance of some scientific discoveries and recognizing their transformative power—especially if this leads to a habit of attending to the best Science journalism available. Beyond that, however, is a fundamental skill: being able to read (Norris and Phillips 2003).

Education in reading does not stop at some early grade. It continues through whatever college or university experience you have and even through your entire life. People learn how to read novels and poems, political articles and analyses, music criticism and ethnographic reports. Scientists learn how to read the specialized articles within their fields. We do not, and should not, expect that level of skill from scientifically literate citizens. It is appropriate, however, to aim at training students, even those who declare that Science is not for them, so that they could read a range of expository literature—articles, books, Internet documents—in which important technical information pertinent to their lives was explained. Citizens do not need to be able to balance chemical equations, or to enumerate the parts of a flower, or to determine the focal length of a lens that is handed to them. They should, however, be sufficiently acquainted with the notion of a gene to laugh when they are asked whether or not they agree with the statement "The difference between genetically modified organisms and ordinary organisms is that the former contain genes and the latter do not" (See Jasanoff [2005] for surprising results about the rates at which some European students are prepared to accept this statement as true.)

Access to Science might be enormously improved if the school curriculum were radically revised. Standard courses with memorization, problems, experiments, and all the other elements needed for training those who will go on to increasingly technical education in the sciences would be important at the beginning. For students whose experiences of these courses convinced them that they did not wish to follow this route, a quite different sequence of courses would be appropriate, one offering some immersion in scientific practice, appreciation of the high points of the sciences (think how art history is taught), and one centered on acquiring major scientific concepts and an ability to deploy them in acquiring information and thinking through serious policy debates. The aim would be to prepare a huge population of happy consumers of the works authored by the successors of Gould and Sagan, Dawkins and both Judsons, Weiner and Greene.

In designing any such curriculum, there is an obvious danger. One must guard against the possibility that technical education in the sciences will be abandoned too quickly, especially if the migration to the "Science for the citizen" track leaves a community of scientists that fails to represent the whole population. The importance of diversity within Science is paramount (as we shall see in the next chapter), and so, too, is retaining those young people who have the talent and the enthusiasm for research. How to encourage those with potential for a happy and productive life in Science to continue technical scientific education is an issue on which there are surely many people more expert than I—possibly even an issue that might be addressed by educational experimentation—but I am optimistic that there is some way of achieving this goal without sacrificing the interests of the many students who, in their adult days, come to think of their Science classes with a yawn—or a shudder. The task of fashioning a thriving research community and a scientifically literate citizenry, with lifelong enthusiasm for learning about new findings, should not be insoluble.

Chapter 8

DIVERSITY AND DISSENT

33. DIVERSITY WITHIN SCIENTIFIC FIELDS

The division of *epistemic* labor advances public knowledge by recognizing possibilities for different groups of people to acquire different types of information to be pooled for common benefit. Previous chapters have been concerned with ways in which this apparently good idea can go awry, perhaps because of doubts about the identity of authoritative sources, perhaps because of lack of access to the information provided. We now turn to a further division of labor, one internal to fields of inquiry. The division of *cognitive* labor obtains when a group of investigators, addressing a common problem, pursue different approaches to that problem. I shall begin by recapitulating ideas I have previously advanced (Kitcher 1990; 1993, chap. 8). This will serve as the basis for a further exploration of difficulties potentially generated from mechanisms that support a valuable division of cognitive labor. We shall be led back to some of the issues considered in earlier chapters and to further possibilities of extending the ideal of well-ordered science.

In everyday life, groups of people encounter problems best tackled by having subgroups pursue different strategies. If the family dog is lost and dusk is approaching, it makes little sense for all those available to search to follow the same route, even if it is the habitual walk. The chances of success are improved, possibly very considerably, if different people go in different directions. It is similar in scientific investigations. The odds of finding the structure of DNA were increased because James Watson and Francis Crick entered the "race" *and because they did something quite different from the careful crystallographic work of their principal "rival," Rosalind Franklin.* (We shall consider later whether this apparently advantageous refinement of the quest for public knowledge has a darker side.) In contemporary molecular biology, cognitive labor is frequently divided in analogous ways: when there are several molecules that might play a role in some developmental or metabolic process, or in the genesis of a disease, research groups focus their attention on different candidates. The subcommunity studying the process or the disease behaves like the family looking for the dog.

It is not hard to idealize these situations to bring out the logical structure behind the familiar conviction that diversity is welcome. Suppose there are well-defined probabilities of success, given an investment of research effort: there is a value, $\Pr(S|n)$, the probability of finding the answer to the problem under investigation by pursuing strategy S if n investigators devote themselves to S. Assume, for any pair of available strategies, the probability that both strategies will deliver the answer is zero: for all values of m and n, and for all strategies S_i and S_j, $\Pr(S_i \,\&\, S_j|\langle m,n\rangle) = 0$ ($\langle m,n\rangle$ obtains when m investigators pursue S_i and n pursue S_j). Given this last condition, the probability of community success given an assignment of n_1 scientists to the strategy S_1, n_2 to S_2, and so on through all the available r strategies, can be expressed as

$$\Pr(\text{success}|\langle n_1,n_2,\ldots,n_r\rangle) = \Pr(S_1|n_1) + \Pr(S_2|n_2) + \ldots + \Pr(S_r|n_r)$$

If there are N investigators within the subcommunity tackling the problem, the community has maximal chances of success when numbers n_1, n_2, ... n_r, whose sum is N, are chosen to make this probability as large as possible. If one strategy, S_1, is much more promising than all the others—that is, for any other strategy S_i, and for any numbers m and n, $\Pr(S_1|m)$ is larger than $\Pr(S_1|m\text{-}n) + \Pr(S_i|n)$—it will be best to put all the eggs in one basket, to have everybody pursue the most promising strategy. On the other hand, if there are diminishing returns to additional investment in any particular strategy; that is, if adding one more scientist to the pursuit of that strategy raises the probability of the strategy's success only slightly, and if the probability of a different strategy's being successful, given pursuit by a single scientist, would exceed that slight amount, it is better to divide the labor. In this particular instance, $\Pr(S_1|m) < \Pr(S_1|m\text{-}1) + \Pr(S_i|1)$, for some alternative strategy i. More generally, if there are diminishing returns, and a large enough group of investigators who might be distributed, it is possible for $\Pr(\text{success}|\langle n_1,n_2,\ldots,n_r\rangle)$ to be maximized if more than one of the n_i is greater than zero.

The formal details sketched here depend on idealizations, and I offer them only to clarify the structure of an important point. We should, I think, be suspicious about the assignment of point probabilities to events like "success-given-that-so-many-investigators-pursue-this-strategy"; the assumption that there is no chance of two different strategies' succeeding is also implausible, as is the (tacit) presupposition that all the workers are interchangeable. These observations do not undermine the value of division of labor. Nobody, no scientist, sociologist, economist, or philosopher can specify the functions that yield probabilities of success given the assignment of *x* researchers—but scientists can, and do, make much rougher judgments. They recognize sometimes that having an extra individual, or an additional team, work on a particular approach is not likely to generate anything new, and that the problem is more likely to be solved if that individual (or that team) does something different. Without supposing they can reach the *optimum* division of cognitive labor, they realize that a *heterogeneous* distribution, the pursuit of different strategies, would be superior to complete uniformity.

When heterogeneity is welcome, how is it achieved? One possibility would be for the members of a research subcommunity to engage in joint planning. They come together and consider the various strategies for solving the focal problem. They share their (rough) judgments about what kinds of distribution of effort would be productive. Having arrived at the division of labor they consider best (so many to this strategy, so many to that), they assign themselves to the various approaches by drawing lots—or maybe they keep records—and follow principles that take into account which lines of pursuit seem most likely to succeed ("You had to work on that really fly-by-night approach last time, so it is only fair that you pursue something more promising for this project"). I know of no subcommunity that organizes its research efforts in this way.

Another option would be for investigators to try to adjust their own work so it is most likely to contribute to public knowledge. Before committing themselves to any specific approach, scientists look around and check on what their colleagues are doing. They make (rough) judgments about which strategies are underrepresented and choose that one they believe will be most responsive to additional effort. Their decisions are thoroughly altruistic: "I intend to devote myself to whatever line of inquiry will add most to the chances of someone's solving the problem." Perhaps considerations of this kind do play a role in the decisions of some scientists. If so, it would be good to identify those scientists and to celebrate them.

Heterogeneity could be produced in a quite different way, one very much in line with the story of the search for the structure of DNA, used above as an illustrative example. Watson's own account of how he planned his investigative strategy hardly portrays him as concerned to maximize the chances that some member of a small subcommunity would solve the problem. He aimed to establish himself as an important scientist and, if possible, to gain the most prestigious honor Science has to offer, a Nobel Prize (1968). Given this aim, he was entirely reasonable to choose the path he did. Had he sought to follow the strategy patiently pursued by Rosalind Franklin, his chances of being the first person to announce the structure of DNA would have been infinitesimal. Franklin had a head start, she was (unlike either Watson or Crick) skilled in the techniques of X-ray crystallography, and it was not obvious how Watson could have argued for access to the necessary equipment. Instead, he and Crick chose the far less reliable strategy of speculative model building, apparently well suited to their background expertise and temperaments, because, even though it seemed much less likely than the crystallographic route to succeed, it offered *them* a better chance of reaching the goal first. Although it was more likely that Franklin's *method* would disclose the structure (or identify it first), it was far less likely that *Watson and Crick* would discover the structure (first) by following that method than that they would make the discovery (first) by speculative model building.

Many years ago, in a Harvard parking lot, the great evolutionary theorist (and historian and philosopher of biology) Ernst Mayr ended a long conversation with me with the cryptic sentence "The trouble with you philosophers is that you think scientists want the *truth*; they don't; they want to be *right*." For a long time, I was puzzled about what Mayr meant. Surely "being right" amounts to arriving at the truth? Light dawned when I recognized that Mayr's formulation involves *scientists* aspiring to a particular attribute, wanting to take on a particular status. They aim at being *the one who found out the truth*. Part of that surely is a matter of personal satisfaction. Even if your status as "the one who found out" is unrecognized, you can still know what you achieved. High-minded detachment can easily be mixed with the desire not only to *be* the one who found out, but to *be known as* the one who did it—to be recognized for the achievement. Prizes and honors are the outward manifestations of that recognition and are coveted as such. Watson's frank account makes that very clear.

Once again, idealizations can help expose the underlying structure of the situation. Suppose, for simplicity, that there are two methods for addressing a particular problem, M and M^*. M is always more promising than M^*, in that the probability that M delivers a solution if n investigators pursue it is larger than the probability that M^* achieves the solution if n researchers follow it: $Pr(M|n) > Pr(M^*|n)$ for every n. Provided there are diminishing returns to adding more inquirers to either method, it is possible that the community's chances of success are greater if some investigators work on M^*: $Pr(success|N) < Pr(success|<N-1,1>)$ (if there are N investigators, the chance of someone's solving the problem if one of them pursues M^* are higher than the chances of someone's solving the problem if they all work on M). Researchers judge the probability of *their* being the one to announce the solution, if they pursue a particular method, by supposing themselves to have equal chances to everyone else who is following that method: in effect, when you pick a method, you buy into a lottery; whether that lottery pays up depends on the chances that the method succeeds, given the number of people who pursue it; if it does pay up, your ticket is as likely as anyone else's to win. Imagine you are the last researcher to make up your mind and that everyone else has chosen M. The probability you win if you opt for M is $Pr(M|N)/N$; the probability you win if you choose M^* is $Pr(M^*|1)/1$ (i.e., $Pr(M^*|1)$). You improve your chances of being the first person to solve the problem by choosing M^*, if $N\ Pr(M^*|1) > Pr(M|N)$, and it is not hard to see how this can occur. (For example: suppose $Pr(M^*|1) = 0.1$; $Pr(M|1) = 0.5$; $Pr(M|10) = 0.9$; and $N = 10$; although M^*'s promise is much inferior, $10\ Pr(M^*|1) = 1 > 0.9 = Pr(M|10)$.)

An apparently *extrinsic* motive, something many commentators have thought would sully the mind of a scientist, can play a *helpful* role in the community search for knowledge. Your desire for recognition might lead you—as it apparently led Watson and Crick—to opt for a less reliable method and thus yield welcome heterogeneity within the group of investigators. As with the earlier invocation of probabilities, it is important to recognize that rough judgments can substitute for the unrealistic idea of assigning exact values to chances and computing the odds of winning. Probably no researcher has ever tried to calculate the most effective strategy for attaining recognition. On the other hand, I suspect that many have deliberated in ways that approximate the contours of the probabilistic relations, operating with a sense that the chances of their solving the problem would be lower if they followed the crowd and selected the most reliable method. In consequence, they have struck out in riskier directions.

The analysis just reviewed, developed in considerably more detail elsewhere (Kitcher 1990; 1993, chap. 8; Brock and Durlauf 1999; Strevens 2003), can be used to defend a number of points. First, there is a difference between what might be reasonable for an isolated individual to choose, to believe, or to do, and what might be reasonable for a group of investigators, facing a common problem; further, the reasonable choices (beliefs, actions) of members of group projects can diverge from those that would have been reasonable had they been isolated agents. Second, there are frequently occasions on which it is reasonable for a community to pursue a number of different approaches, even if one among them seems more promising than the rest (and would have been the reasonable choice for a single individual); diversity is often welcome. Third, diversity can, in principle, be created and sustained in a number of ways: through deliberate planning based on an overview of the options (including options for division of cognitive labor); through individual attempts to appreciate what is collectively best and to adjust behavior to the conduct of others; and through the operation of apparently extrinsic pressures, including some that appeal to motives one might suspect of being antithetical to advancing knowledge. Fourth, the last possibility is worth recognizing when investigations of actual scientific behavior suggest that extrinsic pressures (like national affiliation or the desire for fame) play a role in researchers' decisions: it is not legitimate to draw skeptical conclusions to the effect that the reasonableness or reliability of scientific inquiry is thereby undermined.

My aim in this chapter is to develop the second and third points further. First, I want to think more systematically about what types of diversity within a scientific community, and within the ambient society, might be valuable. In addition, we shall consider the possibility that attempts to sustain diversity might interfere with inquiry in ways earlier discussions have not appreciated. I shall be particularly concerned with the sorts of problems for integrating expertise with democracy that have emerged in previous chapters.

34. VARIETIES OF DIVERSITY

The value of a division of cognitive labor was explained and defended by thinking about a very specific context, one in which a community of researchers faced a common problem for which several different strategies of finding a solution were available. That is by no means the only sort of diversity we can imagine, or even the only sort actually present in the sciences. Investigators within the same field can disagree about which problems are significant, about which standards for certification are appropriate, about what to believe and which people (or instruments, or techniques) to trust. Sometimes there is *radical* diversity, expressed in skepticism about the fundamental approach adopted in a particular field. It is worth asking which of these kinds of diversity might be valuable and which are better avoided. I want to start, however, by looking at a different issue, the potential value of diversity among the kinds of people who enter a research area.

It is well known, and much lamented by the vast majority of scholars, that, for most of the history of most of the sciences, all those involved in inquiry came from a very small subclass of our species. They were male Europeans (or descendants of Europeans) and typically not from the lowest strata of society. At some times and places, they had to satisfy criteria not directly pertinent to the questions they investigated, to know Latin (or English!), to be prepared to profess the articles of a particular religion, and so forth. Many of the restrictions have been abandoned—the Royal Society now admits women fellows, Jews are no longer debarred, tradespeople and their children are admitted—but some endure (we insist on English). Given the educational opportunities available to the world's poor, it is overwhelmingly likely that, although they can participate *de jure*, their representation is, *de facto*, a tiny fraction of the value expected from statistics. Do the past restrictions, or their present residues, matter? Is an increase in the diversity within a scientific community automatically a good thing?

Plainly, there are some characteristics with respect to which homogeneity can be tolerated. If it turned out that people born in March, or those with red hair, or even left-handers or people in the top 5 percent of the distribution for height were dramatically underrepresented among natural scientists, that would be a curiosity but not obviously something calling for reform. Diversity is important when it is supposed that the traits that vary bear on judgments, decisions, and actions affecting the course of inquiry. Given the pervasiveness of value-judgments in scientific practice (§4), excluding people who might bring a different point of view to an ideal conversation is problematic. An instance of the trouble occupied us in considering agenda setting (chapter 5). The actual biomedical research agenda neglects the needs and aspirations of the world's poor, and it is natural to suppose that might be different if people from poor areas of the world—or even people who had spent significant time in those areas—were more common among biomedical investigators. Well-ordered science, in its original form, responds to the difficulty by introducing the ideal of full representation: the perspectives of the human population are to be part of an ideal discussion, and those perspectives must have an impact on any adequate simulation of an ideal discussion. As we have seen, however, some perspectives cannot be introduced into the conversation by those who have them. Nonhuman animals are not able to speak to us, and members of future generations are not here to speak. However we might want to, we cannot include these affected parties in the community of inquirers. Perhaps, then, well-ordered science does not need to insist on a group of scientists that reflects the distribution of socio-economic levels?

In principle, it does not. If, however, as seems plausible, outside representatives are typically less good than insiders at articulating the distinctive needs and predicaments of particular groups, it is worth striving for inclusiveness when there is hope of achieving it (as there is not for the animals or the future people). A principle commending bringing as many different pertinent types within the research community would capture a central theme in democracy (and in its conception of the division of epistemic labor): others may know all sorts of things about the world of which I am ignorant, but I am likely to be a better authority on my own situation and my own aspirations. Hence, rather than thinking of well-ordered science as a surrogate for increasing diversity in the community of investigators, it would be better to envisage expanding diversity as a good way of trying to move closer to well-ordered science.

The benefit of a diverse group of researchers is not restricted to improvements in assessments of scientific significance, however. Different types of people might be inclined to think up different hypotheses or to approach problems in alternative ways. As already noted (§24), primatology was transformed through the increasing presence of women in the field (in both senses). Instead of restricting their attention to the (supposedly determinative) behavior of male members of primate troops (and often only to the "dominant" males), women started to look at what the females were up to and quickly discovered the many subtle ways in which their patterns of behavior shaped male activity (for example by their rewards to "subordinate" males whose support—even friendship—advanced their ends).[1] Much controversy swirls around the possibility that women are "hardwired" to have different cognitive capacities than men. Because the topic has absorbed so many confusions, and because it can be bypassed, it is better to move to firmer ground (Keller 2009). Whether or not the differences would persist in all possible social environments, it is undeniable that, in the environments most women actually experience, they develop to acquire distinctive ways of framing problems and of approaching them. Especially in areas where differences of this sort can be identified, it is well to take advantage of them, but, even where they cannot, the possibility of a valuable expansion of cognitive strategies provides *epistemic* grounds for encouraging diversity.

Parallel considerations apply to any group whose distinctive social embedding (whether or not it is also influenced by supposedly "genetic" differences—a supposition to be treated with profound suspicion) can be expected to make its members sensitive to factors others would ignore. Particularly in areas that might benefit from sensitivities of this sort, diversity of cognitive style is welcome. *In addition to the reasons based on justice and equality for opening opportunities in Science to people who have traditionally been left out, and for ensuring sufficient representation to convince young people that a scientific career represents a genuine possibility for them, the institution of public knowledge can be improved by including as full an array of cognitive talents as can be found.*

Turn now to the questions postponed above, questions about the value of intrafield diversity with respect to judgments of significance, belief, and standards for certification. Two scholars who pioneered philosophical study of the historical development of the sciences, and who drew attention to possibilities of radical diversity, differed profoundly on these questions (Kuhn 1962; Feyerabend 1975; 1978). In Feyerabend's view, diversity and clash of perspectives are always valuable, not only because they are the engines of change, but simply for themselves. Kuhn, by contrast, views diversity as symptomatic of the most primitive stages of scientific inquiry and of those moments when a fruitful consensus has broken down and a field is in "crisis." An emphasis on uniformity might be defended, as we shall see, by appeal to the role experts, including, prominently, scientific experts, are to play in crafting public policies, but Kuhn does not offer this line of defense. His attention is largely restricted to the internal development of a field of science, and he emphasizes the importance of a large basis of shared beliefs and values (cognitive and probative schemes of values, to use our earlier terminology—§4) if problems are to be solved and progress is to be made. Plausible as this thought may seem, there are grounds for thinking that neither of the suggested positions is quite right.

Consider a different view, the *coalescence* model of research. Fields of inquiry constantly undertake new problems, and, for the reasons considered in the previous section, a diversity of approaches is often welcome. During the period of investigation, we want to encourage the clash of perspectives Feyerabend applauds everywhere. More exactly, it would be good to have a distribution of the available researchers among rival strategies that avoided the diminishing returns stemming from duplicated effort. There will be research problems for which complete heterogeneity might prove disastrous, for promising methods might require a critical mass of investigators to have much chance of succeeding. Complete homogeneity is typically bad unless the field is very small, since otherwise there will be enough people to pursue more than one potential approach.

The field will benefit from differences in strategy—but does that require difference in *opinion*? Possibly not. Human psychology might allow dedicated researchers to commit themselves to alternative approaches, and to pursue them energetically, but these are not the figures who emerge from the pages of our historical sources (the journals and letters of past scientists)—nor are they typical of the scientists I know. The diversity of strategies for solving a focal problem is typically accompanied by diversity in belief and often by diversity in assessing the significance of subsidiary problems. Because researchers think they have the key insight, one that their rivals have overlooked, that the questions they address *en route* are the most important ones, they are motivated to spend long hours pursuing a chosen strategy. Whether altruistic investigators who agreed to pursue an approach they thought highly unlikely to succeed, "for the good of the community," would exert comparable effort seems doubtful.

Some variation in standards for accepting conclusions on the basis of evidence is probably also valuable. Within the limits a community sanctions (§24), there will be space for the more cautious and the more impulsive. Evidence that will satisfy the hares will not be enough for the tortoises. Sometimes the hares will race ahead, completing a sequence of studies that culminates in a solution; on other occasions, their conclusions will be premature, and they will come to grief. If the limits accepted by the community are well chosen, there is no general determination of which strategy will prove superior, and it is valuable to have a mix of the two temperaments.

All this applies in the context of solving a problem, but it is tempting to suppose that diversity when the investigation is under way should coalesce into consensus once the problem has been solved. After the fact, investigators who have committed themselves to different beliefs along the way, who have formed alternative judgments about the importance of subsidiary issues, who have evaluated their rivals adversely in light of their own evidential standards ("*X* is leaping to conclusions instead of doing the studies that are needed," "*Y* insists on dotting the i's long after compelling evidence is in") are expected to reach agreement on the solution and to discard any beliefs and evaluations incompatible with it. That frequently occurs. On numerous occasions, scientists publicly admit that they were wrong in the assumptions they previously made, or that they had misidentified some crucial issue: one striking instance is the rapid acceptance of plate tectonics in the 1960s, when prominent researchers confessed to their students that the views they had been presenting were entirely wrong. Some of the negative judgments made in the course of the investigation may persist as grumbles: "This time *X* got lucky," "Of course, that old stick-in-the-mud *Y* could have figured it out more quickly."

The coalescence model fuses the ideas that move Kuhn and Feyerabend by assigning them to different phases of inquiry. When matters are up in the air, when a problem is not yet solved, diversity is welcome, including diversity of opinion, diversity about judgments of significance, and even diversity in standards of certification (within the limits recognized by the community). Once the problem is resolved, the community should unite and build on the achievement; further diversity on the issue would now be wasteful. As we shall see below (§36), it is worth distinguishing two questions about dissent (diversity after an issue has been resolved to the appropriate research community's satisfaction). One concerns the state within the community: is it valuable to have dissenters who resist the conclusions of their colleagues? The other focuses on the broader public: is it valuable for there to be outsiders who dissent?

For the moment, let us restrict attention to the internal development of the research field, asking what distribution of views would be likely to advance areas of science in respect to posing and resolving significant questions. There are two polar proposals. One, the *anarchist* suggestion, denies that consensus is ever a good thing. Anarchists think Science thrives on continual ferment, and it is better if people disagree about anything and everything. The clash of competing ideas produces a richer stock from which to select. The *conservative* suggestion thinks fields need procedures of certification, and, when those are well formulated, and when a problem solution is certified by the procedures, that solution should be adopted by all. The coalescence model has been framed in terms of this conservative suggestion. Because the conservative suggestion is itself an inadequate resolution of an important debate, the coalescence model requires further refinement.

It is easy to understand what conservatives find objectionable about anarchism. The extreme diversity anarchists celebrate provides no basis for the slow and patient cooperative work that underlies so much science. One of Kuhn's great insights (recognized from the beginning by scientists but overlooked by philosophers who gravitated to the provocative theses about revolutionary debates) was his sensitive portrait of normal science, an activity governed by deep consensus on basic beliefs and standards. The anarchist suggestion celebrates the chaos of the "preparadigm" phase—full of excitement but going nowhere. Normal science can tolerate the sort of diversity proposed in the coalescence model and can see it as facilitating the cooperative work of building a picture of some aspect of nature. Yet it depends on the community coming together once the problem solving is done, adding to the stock of knowledge, and using the full resources of that stock in addressing the next series of problems. (As we shall see, this argument about the importance of consensus has a counterpart in thinking about public dissent.)

Nevertheless, the anarchist is responding to something important, indeed to the kinds of considerations moving Milton and Mill (§31). Although one might reasonably worry about the conditions of debate in the public arena, where the vast majority of the participants (or spectators) lack access to crucial items of knowledge and are thus unable to reach any responsible evaluation, internal controversies within a scientific field—held "behind closed doors," as it were—do not suffer from this defect. All the parties can be expected to understand the substance of the debate. Under these circumstances, the benefits adduced by Mill would seem likely to flow from continued discussion: rejection of claims wrongly accepted, refinement of partial truths, deeper understanding of the correctness of items of knowledge. Indeed, the complete consensus envisaged by the coalescence model appears unhealthy, for it is likely to enshrine deep errors, in the way dogmatic systems of belief have so often done.

Conservatives want the community to get on with the further business of problem solving, without wasting time and effort rehashing things that have been completely settled. It is easy to sympathize. Yet it is worth considering whether a complete consensus is required. Apparently, we face a situation analogous to that studied in the previous section. There is an alternative to the research strategy of insisting on (or encouraging) complete consensus. Imagine that the community contains a small proportion of challengers, people who re-open disputes everyone else thinks have been completely settled. Mostly they come up with very little and are ignored. Occasionally, however, they expose some deep presupposition and demonstrate how amending it can open the way to solving problems that have been viewed as significant and difficult. Under some circumstances, this sort of community will do better, from the perspective of advancing knowledge, than one that proceeds by complete consensus.

Once again, the logic of the situation can be exposed by making idealizations. Imagine a research community of size N, and suppose n to be chosen to make n/N small. The expected benefit of the strategy of complete consensus is the sum of the expected benefits of addressing a particular range of problems, supposing everyone in the community accepts all the certified findings. It can be written as

$$\sum \mathrm{Pr}(p_i \text{ is solved} | N \text{ accept}).\mathrm{U}(p_i \text{ is solved})$$

where p_i is the ith problem addressed, $\mathrm{U}(p_i$ is solved) is the benefit of solving it (measured by its significance for the development of the field), and the summation is over the full range of problems. The expected benefit of the alternative strategy is the sum of two terms: one representing the problem-solving benefits, assuming that only N-n researchers are part of the consensus (presumably, since the number of investigators involved in further problem solving is decreased, the probabilities of problem solving go down), and one representing the possible contributions of the n challengers. That benefit can be written as:

$$\sum \mathrm{Pr}(p_i \text{ is solved} | N\text{-}n \text{ accept}).\mathrm{U}(p_i \text{ is solved}) + \mathrm{Pr}(\text{revision} | n).\mathrm{U}(\text{revision})$$

where $\mathrm{Pr}(\text{revision}|n)$ is the probability that the efforts of the challengers will lead to some significant revision (or one of the other good effects Mill takes challenge to orthodoxy to bring) and U(revision) is the benefit of making that revision. If the community is relatively large, the difference in problem-solving success resulting from siphoning off n researchers to a different type of inquiry may be quite small (diminishing returns again!); while the probability of significant revision may be very low (challenges to orthodoxy are usually flawed), the value of making such revision can be large enough to make the second term in the "challengers" strategy greater than the loss in terms of problem solving:

$$\Pr(\text{revision}|n).\text{U}(\text{revision}) > \sum \Pr(p_i \text{ is solved}|N \text{ accept}).\text{U}(p_i \text{ is solved}) - \sum \Pr(p_i \text{ is solved}|N\text{-}n \text{ accept}).\text{U}(p_i \text{ is solved})$$

So there is a real issue about whether complete consensus is valuable.

As usual, the idea of precise probabilities, especially of probabilities someone could identify, is highly unrealistic. Typically, even the most savvy insiders can make only rough judgments. If some researchers spend their time worrying about bits of "established knowledge," that is likely to reduce the rate of further problem solving—but not a lot, if they are a small minority. Their work is not expected to produce much, but if it does, the advance might be far more valuable than many generations worth of problem solving. Perhaps, then, it is good to have a few challengers around. *How these judgments are further articulated, and how the extent of a profitable consensus is assessed, is likely to depend on the details of the field. It should not be supposed that there is a single answer to the question "How complete should the consensus be?"* I conjecture that, in some instances, perhaps in many, the coalescence model will need refinement to allow for the possibility of obtaining the Millian benefits.

As with the division of cognitive labor, it is useful to consider how individual scientists might make their decisions. Crudely, you can think of a researcher as having two possible strategies. If you play *Conformist*, you learn and adopt the consensus within your field, and you try to build on the received ideas to solve outstanding problems. If you play *Maverick*, you pursue some possibility incompatible with orthodoxy in the field. Conformists are relatively likely to succeed in some of their problem solving, to acquire more or less good reputations and honors of various sizes, to enjoy the respect of their peers and a sense of contributing to a valuable cooperative effort. Mavericks, on the other hand, may beat their heads against the wall for years, what results they take themselves to have achieved may be ignored or dismissed by others, they may be written off as loons (or worse); set against that is the dim possibility that they will articulate something truly novel, that it will ultimately be adopted by the research community, and that they will go down in the history books among the great figures—the truly great figures, not "merely" the Nobel Laureates—in the history of inquiry. Mavericks play for very high stakes, with a tiny chance of success.

How scientists choose among these options is surely a matter of temperament. Many are inclined to value the solid benefits of conformism, to appraise their own powers of imagination and depth of thought soberly, and to see life as a maverick as not for them. Others are stirred by ambition, confident of their talents, captivated by a particular idea, and robust enough to withstand years of neglect and scorn. Fame is the spur . . .

The categories just introduced are admittedly crude. You might worry about how they actually apply to the history of the sciences (Was Newton a maverick?). More importantly, a binary division ignores the different levels at which orthodoxy can be resisted. Some mavericks want to reconceive a field completely; others maintain part of the consensus while challenging other elements. More limited challenges might well increase the chances of success, but they bring smaller rewards. Thus we might expect differences in temperament to translate into divergences at different levels (or at different scales) for a relatively large research community to contain a mix of types.

Would the simple pressures to which I have pointed generate a satisfactory (even an optimal) pattern of consensus and challenge? The question is unanswerable, for, as noted earlier, there is no general determination of the amount of deviation that would be valuable. Nor can anyone know all the factors potentially inclining researchers to choose whatever role they find most apt for themselves. Failure to resolve these issues does not make reflection on diversity, its value and its potential sources, otiose, but rather it encourages a more pragmatic attitude. We have recognized the possibility of promoting fields of research by including a small population of challengers, as well as noting the influence of modesty and ambition on producing a mix of types. Without undertaking the impossible project of a general analysis, these considerations can be applied locally in reflection on the state of a field. A further component of well-ordered science is its maintenance of a healthy level of diversity, where the standard of health depends crucially on the achievements and the difficulties of particular areas of inquiry. Judgments as to whether some domain is too hidebound, too reluctant to explore quite fundamental ways of revising its orthodoxy, or whether it is too open to speculative challenges, may inevitably be rough, but they are neither impossible nor unimportant. In accordance with the ideal of well-ordered science articulated earlier, they are best made through deliberations involving a wide range of perspectives, undertaken by discussants who are tutored and committed to mutual engagement. The concrete proposals for moving in the direction of the ideal can be further elaborated by a slight extension of the mandate of those groups that are led "behind the scenes," and whose verdicts address the problems considered in earlier chapters: part of their mission is to reflect on various kinds of diversity within the fields they study, and to offer conclusions about whether more or less diversity might be appropriate and about how any necessary changes might be made.

35. MARKETS, NORMS, AND TEAM PLAYERS

Suppose it were to be a good thing, and to be recognized as a good thing, that the community of inquirers, either within a particular domain or in its entirety, should have a particular feature, one that depended on the coordinated efforts of a number of individuals. The coordination needed could be brought about in several ways. Here are three:

> *Cultivating team players.* Coordination results from the creation of a social environment in which the parties develop a strong tendency to recognize the valuable feature and to act in ways that, given the choices of those around them, bring it about.
>
> *Instituting norms.* Coordination results from the presence of norms, universally recognized in the community, that direct the actions of individuals in ways leading to the presence of the feature.
>
> *Market forces.* Coordination results from the individual decisions of the parties, who seek to promote their own ends; given the structure of the environment, their actions lead to the presence of the feature.

Many analysts of scientific practice write as if they emphasized one of these as the *sole* determinant of the features they explain. (Robert Merton is, for example, an obvious fan of norms.) I shall argue against a monothematic approach: research and public knowledge thrive best when the three potential mechanisms buffer, reinforce, and sometimes limit one another.[2]

For an example of explanations that invoke one of the mechanisms, consider scientific fraud. Until quite recently, many commentators on Science were struck by the thought that fraud is relatively rare (suspend judgment about whether or not they were correct in thinking this). Many commentators who subscribed to the Galtonian image of scientists as forming a secular priesthood were inclined to ascribe this to the honesty of individual researchers (and this account lingers in contemporary journalism about Science). Perhaps the idea is that scientific research attracts particularly high-minded people; or maybe the training people undergo in becoming scientists inclines them to be more virtuous than they would otherwise have been—they absorb honesty at the bench. By contrast, Robert Merton explicitly rejected the thought that scientists are people with special qualities of character (1968). On his account, scientific practice is dominated by a public norm, one that practitioners learn and learn to think of as important. They shun fraud not because of their virtue—perhaps outside the lab they cheat in all kinds of ways—but because, in this area of their lives, they are completely accustomed to thinking of honesty as paramount. Finally, you might think scientists are neither particularly virtuous nor especially inclined to obey abstract norms. Cynically, you see them as much like other people: willing to cheat when it suits them and to break the rules if they believe they can get away with it. What holds them in check is the structure of the social environment in which they work. Science is partially cooperative, partially competitive. Because of the cooperative aspects, others will try to build on your contributions; because of the competitive facets, they will suspect you if their attempts to build go awry, and, if they become convinced of your deceptive practices, they will not hesitate to expose you. Hence it is in your interest to play it straight.

None of these explanations is completely convincing, and, as more becomes known about different episodes of scientific fraud, there are grounds for wondering about the supposed phenomenon to be explained. *Is* scientific fraud rare? Do we have any good basis for estimating its incidence (§23)? The explanations are useful primarily because they point toward measures that could be taken to reduce the incidence of scientific fraud (whatever it is). One possibility would be to try to fill the scientific community with especially honest people. That might be done by using probity as a ground for admission to specialist training (Why are some traits of character thought irrelevant in assessing candidates, even though they might affect performance in the role those accepted will eventually fill?), or it might be done by insisting on extreme candor during the scientific apprenticeship (think of the ways parents try to imbue their children with a horror of lying). Another possibility would be to emphasize, again and again, some clearly articulated norm, forbidding particular kinds of practice: scientists are required to keep complete records, to show them on demand, to explain to designated mentors the steps they have taken to ensure that the reports they issue correspond to what has been done. Yet another would be to structure the environment to raise the chances of catching those who cheat and to make the consequences truly painful. It would be odd to think of pursuing one of these approaches to the exclusion of the others. Why not use all the tools available?

I have returned to a discussion of fraud simply to introduce the idea of a polythematic strategy for creating and maintaining valuable features of a scientific community. My principal concern, however, is with the specific feature of diversity, or, more exactly, with the types of diversity considered in the two previous sections. Here, there is an important asymmetry among the mechanisms. Perhaps a few researchers are real team players, asking themselves from time to time how the work they do contributes to the broader enterprise of promoting public knowledge. Unless my experiences of scientific practice are distorted, however, reflections of this sort are relatively rare, and the disposition to direct one's research where it could do most to advance some general good is even less common—this is not to criticize scientists, for similar self-scrutiny and altruistic concern is rare across the board (including among philosophers!). At best, *team playing* is a very weak force in producing diversity.[3]

Nor can much be expected of norms. There are no widely recognized norms enjoining researchers to act in ways that would produce various types of cognitive diversity, except those that do so indirectly through urging the inclusion of different kinds of people (women, members of minorities, possibly people from very poor societies). With respect to many of the sorts of diversity previously singled out as valuable, it would be very hard to institute norms directly generating the desired conditions. As we have seen, for many of the predicaments that demand diversity, it is impossible to specify in general the extent or nature of the appropriate diversity. "How many mavericks should there be?"—that depends on lots of things, including the state of the field and, as the next section will argue, the needs of the public. Norms function well when they are easy to apply, within the circumstances experienced by those people whose conduct the norms are to govern. A norm enjoining investigators to promote the diversity of approaches, beliefs, judgments of significance, or standards for certification would be too vague. Anything more specific could not be relied on to yield valuable outcomes.

Market forces, however, can promote recognizably valuable types of diversity. That was evident in the context of problem solving (§33) and in the fostering of a small class of challengers, at least in the crudest and simplest form (§34). Two difficulties remain. One is that of finding ways of generating and maintaining valuable diversity in contexts where the market mechanisms I have described do not apply, *through any of the three strategies that proves usable, and by deploying as many of them as possible*. Second is that of coming to terms with the *damaging* effects of the mechanisms actually promoting division of cognitive labor (and challenges to orthodoxy). Overcoming these difficulties requires the polythematic approach recommended earlier.

Start with the issue of the deleterious side consequences of inspiring scientists to compete for credit (and of instituting a winner-take-all regime; Strevens 2003). To fix ideas, recall some aspects of the episode used to illustrate the operation of market forces, the search for the structure of DNA—aspects not dwelt on in my original presentation. Rosalind Franklin was loath to share her X-ray diffraction photographs with Jim Watson, but Watson acquired some information by means commentators have sometimes not found entirely praiseworthy (Watson 1968; Sayre 1975; Olby 1974; Judson 1979). Watson's own self-description candidly portrays him not simply as a competitor but as a *ruthless competitor*, willing to find out what he could about his main rival's progress and unwilling to divulge anything that might advance anyone else's program. Perhaps all's fair in love, war, and science, too—and those who hold back are overnice, even squeamish, as absurd as people who insist on the etiquette of cricket past. Yet whatever the verdict on Watson's behavior (and historians have differed), his self-portrait raises an important general question: how does the promotion of cognitive diversity through competition for credit affect other desirable features of Science, specifically those depending on cooperation among investigators?

The example just considered points to one form of cooperation potentially undermined by no-holds-barred competition for credit. A community's efforts at solving a problem would often be advanced if different investigators (or teams) pursued alternative strategies (as §33 suggested) *and* if, along the way, the researchers shared information about what they had discovered. Promoting diversity through competition for credit seems likely to interfere with information sharing: the Watson approach—find out as much as you can about what your competitors are doing without divulging anything about your own progress—looks like the most promising strategy. Suppose there is a preliminary problem that must be solved by any of the alternative approaches to the focal problem. Your lab devotes months of hard work to ingenious attempts to solve the preliminary difficulty, and you are finally successful. You announce your conclusions to your rivals or make your materials available to them. Although their efforts have hitherto been stymied because they had not solved the preliminary problem (perhaps they had not even recognized it), they can now quickly take advantage of what you have provided, build on techniques or experiments they have been fashioning during the period through which you were working toward your solution, and announce a solution to the focal problem. Even though they play fair with you, sharing their techniques or communicating their results, your concentration on the preliminary phase has left you in a state where you are less familiar with the steps that now need to be taken. Once your achievement is public in the community, you are no longer competitive.

Under a winner-take-all system, the emphasis on competition thus seems hostile to information sharing. Yet there are advantages to sharing *some* information. Casual discussions with competitors can acquaint all the researchers with a general idea of the strategies pursued within the community, and, as individual labs start to achieve results on preliminary problems, they can start to see possibilities of collaboration with others. You have discovered something or developed a technique of production that might usefully be integrated with someone working along different lines: if the two of you team up, your chances of winning the competition are increased (although you will have to share the glory with your new ally). Competition need not confine information completely, but it is likely to induce patterns of information flow that are partial and opportunistic. Coalitions may form, and the process of coalition building may escalate, but the networks along which new knowledge travels will not extend to the whole community.[4]

Simply instituting a norm requiring disclosure of partial results to all members of the research community will surely be ineffective: although there are circumstances under which it might be to your interest to form alliances and to share with your allies, revelation to everyone is almost always inimical to the search for credit. Moreover, you will always be able to find reasons for defending your reluctance to broadcast your achievements—"We didn't want to announce it until we were completely sure all the glitches had been straightened out." A more appropriate solution would be to regulate the market for assigning credit, so it honors people who have made important contributions.

Actual coalitions emerge within research communities when scientists learn—often in informal settings where they relax and gossip—that rivals are working along lines that might fruitfully be combined with their own. The erstwhile rivals become future allies, and, if they succeed, they share the credit. If information flowed freely and openly in all directions, the solution would probably be announced by an *actual* coalition that left out some people whose work was critical to achieving it—that is another way of expressing the conflict between competition and cooperative sharing. The winner-take-all system assigns credit to the *actual* coalition (or individual) making the announcement. Yet why should the fair assignment of rewards be hostage to the contingencies of *actual* coalition formation, to the chance encounters in the bar and the after-hours gossip? The unlucky scientists, or teams, whose shared results are used by the victors may not belong to the *actual* winning coalition, but they belong to a *virtual* coalition that has produced the solution. A system of giving credit to the *virtual* coalition—and apportioning it by the size of the contribution—would encourage sharing of information. Norms for assigning credit in this way are more likely to be observed than simple directives to share, and they would regulate the "credit market."

The proposal extends an approach to scientific achievement that sometimes figures in contemporary scientific research. Nancy Wexler is well known for her pioneering efforts in discovering the genetic basis of Huntington's disease, as well as for her leadership in raising and addressing ethical and social issues surrounding human genomics. With her father, Milton Wexler, and her sister, Alice Wexler, she has also been part of an institution that might serve to illustrate a less competitive scientific ethos. The Hereditary Disease Foundation supports biomedical research but requires the scientists whose investigations are funded by it to share their results and research products with one another. To quote the foundation's website: "The

HDF has changed the way science is done. It has made collaboration and knowledge-sharing the guiding force behind research." The Wexlers effectively invite researchers to think of themselves as part of a team, dedicated to solving important problems—together.

Restricting information flow is not the only way in which an emphasis on competition has unfortunate consequences. The credit system easily distorts the division of epistemic labor. That can occur through the acquisition of an entirely unjustified aura of expertise. The honors publicly bestowed on researchers who make some heralded discovery often provide them with opportunities for speaking out on a range of issues far beyond those to which they have devoted their distinguished careers, and secure for them a respectful, even deferential audience. Although many great scientists remain modest and cautious, unwilling to claim authority in areas beyond the specialties in which they have succeeded so brilliantly, this is by no means always the case. When others think of you as "Very Smart" (period), it is quite easy for you to think they must be on to something, and to offer your advice on any and every subject under the sun. Even if you talk rubbish, you will receive respect and credence, and that can interfere with the public understanding of important issues and debates (Oreskes and Conway 2010).

A related way in which the competition for credit distorts the division of epistemic labor involves the many researchers who are less successful than the brilliant few who feel the temptation to generalized punditry. In some areas of research, particularly in the human sciences, an approach whose apparent promise excites its champions may be viewed more cautiously, even with suspicion and disdain, by partisans of other visions of the field. The enthusiasts receive limited credit *within* the field. If, however, the claims they advance resonate with popular ideas about human nature, especially if those ideas are no longer "officially" well established, there are possibilities for gaining a different type of credit. They can "go public," announcing themselves as innovators whose bold insights are in advance of the hidebound assumptions of their peers. The praiseworthy trend to encourage presentations making scientific ideas more broadly accessible (§32) provides new avenues for obtaining credit, encouraging the enthusiasts to present their ideas "to the people." Because they dare to utter and defend "truths we are not permitted to express," they find a ready audience, and their claims to provide compelling evidence can easily distort public debate. The popularity of the crudest forms of evolutionary psychology, evident in "explanations" of rape, child abuse, and "mate choice," is a prime instance of this phenomenon (Vickers and Kitcher 2002; Haufe forthcoming).[5]

A third place affected by the emphasis on competition is the conduct of research. If you are in a race to solve a problem, it is crucial for you to get results quickly—and one way to speed things up is to cut out some of the work. You will feel pressure to reduce your samples to the minimum size to provide grounds for going on to the next phase or perhaps even below that size. In principle, you ought to do an extra study, using a different molecular trigger or a different experimental organism, but you feel confident about the way it would come out. So, wanting to make advances on the large problem as quickly as possible, you decide to omit the study. Unfortunately, you know others will expect you to have done it, and they will be critical if they know you have not. How to solve your problem? Through fraud, of course. You claim to have data of the sort expected. *In extremis*, if challenged to produce the data, you can always go back and "redo" the (nonexistent) study.

Whether things like this happen frequently, or happen much at all, the emphasis on competition for credit encourages sloppy, or even nefarious, procedures. We can hope that the norms of research are sufficiently powerful to blunt the temptation to cut corners—or even that investigators are people of such virtue that temptation is never felt. Yet it would be well to supplement our hopes with measures that recognize the dangers and try to address them. Our resources divide into the three types originally distinguished: encouraging team players, instituting norms, and taking advantage of market forces. I have been arguing for the need to integrate all three, for the "market" clearly needs regulating.

Competition for credit generates some welcome kinds of diversity: the division of cognitive labor is often a good thing, and it is useful to have a few challengers and mavericks among us. The existing structure of the credit market threatens valuable practices of information sharing. That difficulty could be (at least partially) resolved by adjusting the market to distribute the rewards in a different way—to attend to the virtual coalition, rather than the actual one. It is less easy to see how market adjustments can help with the problems of Universal Punditry or of marketing controversial defenses of widespread prejudices as if they were "scientific" justifications of bold new insights, for the perception of public acclaim as an important form of scientific credit plays a valuable role in opening up access to Science. In the case of fraud, the principal market source of disincentives lies in the consequences of exposure. Exposure requires detection, and, unfortunately, where fraud has been detected, a retrospective appraisal reveals how the perpetrators could easily have been far more cunning. Sir Cyril Burt might perhaps be excused for thinking that he could make up samples with exactly the same statistical properties, but how, after Leon Kamin's famous unmasking of Burt, and in the age of personal computers, could Robert Slutsky have tried the same naive trick? (Kamin 1974; Engler et al. 1988).

Even if we could aspire to adjust the markets to avoid the difficulties flowing from intense competition for credit, it is clear that there are valuable sorts of diversity not to be attained by regulation of this sort. Recall that there are no general specifications of the extent and types of diversity appropriate for research-in-the-abstract. Nor is it even possible to list a few parameters on which those specifications might depend. Adjustment of investigations to achieve diversity are inevitably local. It is impossible to institute norms that would direct optimal—or even satisfactory—distributions.

What would be valuable, I suggest, are norms that might produce beneficial results less directly, norms requiring periodic reflection on the state of inquiry and one's contribution to it. Most of us find it comfortable to operate by habit, to take it for granted that the role we play, day by day, needs no adjustment or even radical revision. Scientific inquiry is especially vulnerable to the assumption that reflection on one's plans is not required, precisely because of the widespread image of the autonomous investigator. Chapter 4 viewed that image as a residue from a different age, one in which the nascent sciences had a very different relation to public knowledge than the one their developed versions enjoy today. Chapter 5 argued against just the type of autonomy in question here, an attitude seeing decisions of what problems to address as answerable to no one. Well-ordered science might be further extended to demand reflection by specialist inquirers and by the groups of tutored outsiders who mediate between them and the wider public, not only on what problems might be tackled but also on the possible mix of strategies that might be valuable. Consideration of whether a field is in danger of stagnating because too little is done to challenge fundamental assumptions or to try out unorthodox methods—or, conversely, whether research is too scattered—might become a staple of intrascientific discussion. Even if investigators do not think of themselves as good candidates for striking out in novel directions, a sense of the need for greater diversity could be expressed in advice to colleagues and students.

In one of his modes of thinking about morality, Kant invited his readers to consider themselves as lawmakers in an abstract republic, a kingdom of ends. The ethical perspective offered in chapter 2 is at a far remove from Kant's, but I want to complete my account of well-ordered science by proposing a similar idea. Under well-ordered science, individual investigators think of themselves as citizens of an ideal republic of knowledge, with responsibilities to do what they can to promote the good of advancing public knowledge along the lines where it is needed (lines determined by the ideal deliberation that is the origin of well-ordered science). Those responsibilities include periodic reflection on their actual and potential contributions, informed by their understanding of the current state of the area(s) in which they work. The mechanisms of instituting norms and altruistic participation in a collective enterprise fuse in the idea of an ethos of scientists as team players.

I began with the thought, familiar among contemporary commentators on Science, that attractive features of scientific research are probably not best explained by supposing scientists to be an especially high-minded lot. If the form of high-mindedness in question is a commitment to team playing, it is quite likely that insistence on the autonomy of the researcher diminishes the frequency with which it is found. Central to this book is an argument that that insistence on autonomy should go, to be replaced by a very different image, elaborated in the previous chapters. Were that image to become widely accepted, it could generate the ethos I am recommending, for similar self-images routinely do so in other areas. Perhaps thinking of oneself as a member of a large and important team might play the role, more concretely and more realistically, that Galton intended to inspire with his talk of a "secular priesthood"?

36. DISSENT

Diversity before matters are settled, or settled according to the standards of certification of a community, corresponds to dissent afterward. Some systems of public knowledge have brooked no dissent: it is important that all members of the group in which it is transmitted accept the lore of the ancestors or the dogmas of the true church. Section 34 considered the (Kuhnian) worry that dissent *within* the scientific community might interfere with the efficient growth of knowledge. I shall close this chapter by considering a different form of dissent, one marked by the existence of a large proportion of outside citizens who challenge a consensus within the scientific community.

More exactly, suppose the entire research community officially accepts a division of epistemic labor, according to which a particular subcommunity is authorized to certify a particular claim about some aspect of nature. *Extreme* dissent occurs when all members of that subcommunity agree in certifying the claim, no members of any other scientific subcommunity disagree, but a significant fraction of outsiders in the ambient society challenge the claim. *Amateur* dissent occurs when all members of the subcommunity agree in certifying the claim, some members of other research communities challenge the claim, and a significant fraction of outsiders does so too. *Minority* dissent occurs when a large majority of the subcommunity certifies the claim, but a minority does not; again, a significant number of outsiders challenge the claim. The notion of significance that figures here is tied to the formulation and implementation of policies. If a *significant* fraction disagrees, it becomes difficult to craft policies based on the claim and to put any such policies into effect.

These specifications allow for a development of the Kuhnian point within the public political context. Dissent holds up concerted action (not merely the advancement of knowledge), and that can be disastrous if there are problems for which policies are urgently needed. In the case of intrascientific disagreement (§34), it was possible to point to a potentially countervailing benefit of continued challenges, the possibility mavericks might uncover something fundamentally wrong with the consensus. Can anything similar arise here?

Suppose the subcommunity has proceeded as well as you could hope. Its procedures for certification are reliable, and they have been responsibly applied. If the case is one of amateur dissent, efforts have been made to engage with the ideas of the nonspecialist scientists: their doubts and arguments have been thoroughly heard. If it is an episode of minority dissent, similar consideration has been given to the minority of insiders who challenge the majority view. Perhaps it turns out that the divergence of opinion turns on different probative schemes of values. If so, the differences have been thoroughly aired, through the frank explanation of the considerations behind the various schemes. They have not led the majority to change their minds.

Under such circumstances, is there a role for the public to play, one analogous to that of the hypothetical challenges of §34? The obvious suggestion is that, for all their efforts, the scientists might be wrong. Even in extreme dissent, the scientific judgment might err. Of course. But if you ask whether a group of well-trained researchers, thoroughly familiar with the details of the issues, proceeding in the patient—indeed, ideal—way described will be *more likely* to be right than an uninformed public, the answer seems obvious: even if you cannot be sure, you know where to place your bets. The obviousness of the answer decreases only a little if the situation is one of amateur dissent rather than extreme dissent, and only a little more if it consists of minority dissent.

An obvious proposal: in any of the three cases of dissent, the public divergence is to be ignored and policies crafted and implemented, no matter what resistance the citizens offer. (You might want to restrict the recommendation to extreme dissent and amateur dissent, or even to extreme dissent.) If the scientific community has lived up to the highest standards, it is a consequence of the division of epistemic labor that its judgments should be used to guide public action. I shall not defend this proposal, but it is worth seeing first what can be said in its favor.

What could account for the public dissent? It is easy to imagine unsavory scenarios. Commitment to chimeric epistemology is playing an unfortunate part. The public is being duped by Universal Punditry: amateur dissent arises because people who have (rightly) been honored for their scientific work make pronouncements on issues about which they know very little. A minority of insiders within the field, competing for credit, go to the public and foment opposition. Perhaps the people who beguile the public, whether researchers within the field or scientists from other areas, are moved by values about which the citizens know nothing, goals that would appall them if they were to find out: perhaps they are in the pay of Big Oil or Big Pharma or Big Tobacco (in a situation in which the Bigs want to thwart the policies they envisage; Oreskes and Conway 2010). As you reflect on these scenarios, it becomes evident that the dissenting public is not simply being unreasonable (putting its unreliable judgments ahead of those of people it has no reason to distrust) but also hoodwinked. The case is like that envisaged by Rousseau in his account of the authority of the General Will: the preferences expressed by the dissenting citizens diverge from their real interests, and the gap is generated by a type of deception. Shouldn't they be "forced to be free," to go along with what they really wanted?

The policy *is* coercive, and it even smacks of the elitist coercion central to Plato's *kallipolis* (§2). The mere fact of the research community's having proceeded impeccably does not entail that the citizens *know* this to be the case. Past interactions between Science and public policy may have generated a condition of alienation (§31), in which, rather than thinking of the minority or amateur spokesmen as being in the pockets of special interests, the citizens suspect that the scientific consensus hides a secret agenda.

A contrary policy, one of deferring to the public, would be to substitute vulgar democracy for the real thing. As in previous discussions, my approach here will embody the thought that public voices need both to be heard and to be tutored. The Ideal of Transparency (§24) is important. How exactly is it to be implemented?

I recommend doing exactly what the scientific community (and its allies in other areas of research) has actually done in instances like this; to wit, attempting to explain, as clearly, completely, accessibly, and patiently as possible, the basis for the consensus. (A famous example is the response to concerns about gene splicing that culminated in the famous Asilomar Declaration.) That will sometimes entail going further than scientific spokespeople have actually done and exploring the values underlying particular decisions (Keller forthcoming). If the counterarguments of dissenters are explicitly considered, and if replies are offered—replies that succeed by the standards not only of the scientific community but by those adopted by a group of well-tutored outsiders with diverse background perspectives and committed to deliberating under mutual engagement, an important responsibility has been discharged: that of communicating fully and openly with the public (ideally the human population) the research community is supposed to serve. To do that is the expression of a commitment to democracy, and, after doing that, no more can be done.

But what if nothing happens? The spokesmen speak with the tongues of men and of angels, and some fraction of the dissenting citizens hear them (or read them). Dissent lessens a bit, but not enough to make the crafting and implementing of policy go smoothly. How should the scientific community respond to immovable dissent?

We have seen that the ideal of free debate, eloquently expressed by Milton and Mill, cannot demand open discussion when the arena is distorted—and, in our times, the arena is *seriously* distorted. (As I write this chapter, news sources report that 20 percent of the American people believe Barack Obama is a Muslim. For fifty million people—at least, for it would be more if the *majority* view were wrong—to be in error about so simple a matter of fact requires real deficiencies in the channels of transmission.) Sections 30–32 offered tentative suggestions about how to improve the situation, but what is the right course to be taken in the interim? I now want to offer a second thesis about the arena in which "free debate" is to occur: not only is it distorted, but it is easy for it to become overcrowded.

If it is good, as Milton and Mill suppose, for ideas to be freely and openly discussed, it is also true that human capacities for attention are limited. (Recall the estimates offered by Robert Dahl, §12.) Questions about which topics are worthy of attention, and about how frequently they should be aired, are exactly parallel to the issues about which lines of inquiry should be pursued, the original scope of well-ordered science (chapter 5). Just as vulgar democracy would give the untutored majority sway in the determination of the course of research, so, too, a supposedly democratic proposal to leave the public arena to the voices that shout the loudest and demand constantly to be heard would not be an expression of the deepest democratic ideals. In many contexts we are quite used to recognizing that decisions have to be made as to how the existing airtime is to be fairly partitioned, but the largest question about free debate—"Who sets the overall agenda?"—is left to the clamor of contending voices, often stimulated by the money of small interest groups.

So arises a radical misshaping of public debate. It is possible for the same controversy to be raised, in the same terms, with the same arguments and the same resolution innumerable times. The public arena does not function as a court in which ideas can be tried fairly and disputes can be resolved but in which it is not permitted to bring the same case twice. Dissenters who fail today can try again tomorrow—without bringing anything new to the discussion. When this occurs, pious appeals to the importance of free debate are actually frivolous: *for if free exchange of ideas has previously taken place, issuing a verdict the dissenters dislike, to bring precisely the same arguments forward again shows total disrespect for the value of the exchange*. Milton, Mill, and their successors saw free and open debate as the occasion for attending to considerations that had not previously been carefully weighed. They would have been appalled at the debasement of their ideal to a series of exercises in repetition, in which dissenters attempt to achieve the verdict they want by wearing down their opponents.

Well-ordered science supplies a model for the shaping of fruitful dissent.

Ideally, the debates that go on in the public arena should focus on significant topics with respect to which there are new things to be heard. Judgments of significance and of evidential novelty can best be made in conversation under the conditions of mutual engagement for which the best approximations are those small groups of well-tutored citizens representing a diverse range of perspectives, the bodies of outsiders who mediate between Science and the public. In this context, they function as grand juries, charged with the task of determining whether a particular style of dissent deserves public attention. When that dissent has, in their judgments, been fairly heard on previous occasions, when the latest expression merely repeats what was put forward in the last hearing, they should declare that the dissenting voices are unworthy of further attention.

How could any such declaration be effective? The dissenting voices may still shout, and their monied backers may supply them with megaphones. Broadcasting their challenges is relatively easy, given the fractured state of the media and the economic forces driving media competition. Yet there is at least one way to give power to the verdicts of citizen juries. Unless dissenters have presented a convincing case that there are new grounds on which the debate should be reopened, any publication of the dissent must carry a preliminary statement to this effect. The vigorous attack on orthodoxy must be prefaced with something analogous to the warning labels on packets of cigarettes (perhaps in the more straightforward style found in the United Kingdom—"Smoking kills" is more honest than "Use of tobacco may be injurious to your health"): "This is the same old story, refuted on previous occasions" or "The council of impartial citizens has not been given any evidence to think that the following discussion contains anything not already considered in discussions of this topic." The analogy is appropriate, for, if free debate promotes intellectual health, it does so only when the public arena is not abused. Part of the task of regulating that arena consists in issuing licenses to those who are serious and thus distinguishing them from frivolous intruders who substitute dogma for discussion.

37. POLITICAL ENTANGLEMENTS?

Although my principal purpose has been to articulate an ideal for the relations between Science and democratic values, previous chapters have made some—tentative—proposals as to how that ideal might be better approximated. These proposals all embody the thought of small, but representative groups of citizens being tutored in particular aspects of scientific research and then deliberating with one another about what courses of action would be most beneficial for all. As I have sometimes noted, there is an obvious worry that processes of this sort will fail to fulfill the intended function of mediating between the community of investigators and the broader public, particularly in restoring public trust, because they will be viewed as politically entangled. Simply put, the citizens who return from their tour of some research area will appear no more trustworthy than the experts who tutor them. Those who oppose the research consensus—people in conditions of resistance or alienation—will dismiss these "citizen representatives," holding them to have been brainwashed.

The concern can be sharply formulated as a dilemma. For the bodies of citizens to serve as credible witnesses, the form of education they have received must be perceived as free of bias. So if they are tutored through being provided with expert accounts of the consensus and the evidence for it, those of their fellow citizens who suspect that expertise has been falsely attributed will reasonably be skeptical of the fairness and balance of the process. If, on the other hand, the trip "behind the scenes" involves genuine confrontation with the full range of perspectives, offering debates in which challenges to the consensus are properly and fully posed, it will simply replicate the confusion that would attend "free discussion" before the citizens as a whole. Either the citizens are turned into replicas of the "expert" scientists—in which case they cannot return as credible witnesses—or they continue to reflect the public confusion—and thus fail to clarify the situation.

My original thoughts about well-ordered science and the potential of groups of citizens to participate in deliberations that are simultaneously broadly representative and well-informed were advanced in ignorance of the actual experiments that have been carried out. Since then, I have learned of two major attempts to realize deliberative democracy.[6]

One of these, that carried out by the Jefferson Project, has sometimes addressed the kinds of questions with which I have been concerned: it has, for example, organized a "citizen jury" on the topic of climate change. The other, pioneered by James Fishkin, has carried out deliberative polling on a number of important political issues, at a number of scales, in a number of countries, and has documented some very impressive results (2009). Yet the most successful instances of deliberative polling focus on topics in which people's knowledge can be increased without raising serious skepticism about the quality of the information presented. When the concern is to arrive at views on the proper extent of US aid to foreign nations, and the necessary education consists in explaining the structure of the budget, suspicions about brainwashing can be allayed.

Of these two models, the deliberative polling of Fishkin appears better suited to the functions I have in mind. The citizen juries of the Jefferson Project are typically too small to count as representative in the way approximations to well-ordered science would require. More crucially, the processes they institute, exemplified in the case of the climate-change jury, recapitulate the adversarial elements in public discussion: spokesmen for different perspectives on climate change were invited to address the jurors, effectively creating a series of disjoint sessions in which radically contradictory views were propounded (Jefferson Project Report). The citizen juries thus lack resources for advancing beyond the failures of "free discussion," diagnosed in §30. Fishkin's deliberative polls, by contrast, involve a larger group of citizens (with sophisticated devices for promoting representativeness) and centrally involve the presentation of information—the kind of tutoring well-ordered science envisages. The challenge is to adapt a structure like this to contexts in which the facts of the matter are themselves in dispute.

How might that be done? What seems to be required is a consideration of alternative points of view, challenges to orthodoxy, in a context in which those who are to evaluate the research topic are closer to meeting the conditions under which free discussion would succeed. Imagine, then, a three-stage process. At the first stage, the scientific community concerned elaborates central concepts and standards of evidence that function in reaching their consensus. In the climate change case, the citizens would learn about measuring temperatures for preceding centuries and millennia by using proxies, and they would be informed about the qualities that make for good proxies; they would also be told about the statistical techniques used to present trends, and their virtues and vices. The second stage of the process would focus on challenges to consensus. This might well involve inviting the most sophisticated spokesmen for the opposition to articulate their various charges. With respect to climate change, citizens would learn about alternatives to the "hockey stick" picture of the history of our planet's mean temperatures, and about the considerations opponents of a recent rise take to favor their rival analyses. The third and final stage would consist in attempts on the part of the scientific community to respond to the counterarguments, explaining to the citizens, *in ways they could be confident they understood*, the reasons why the opponents are mistaken. At this final stage, the citizens would be encouraged to press until they had resolved all their doubts. Only when—and if—full resolution was achieved would the citizens be prepared to report a unanimous verdict on behalf of the scientific consensus.

All stages of the process would embody the interactive discussions so finely articulated in Fishkin's approach to deliberative polling (2009). On complex questions, the entire deliberation might take considerable time—just as jury trials in some cases take a long time. The hope would be that the thoroughness of the investigation would provide the eventual verdicts with sufficient credibility to reinforce public trust—where trust is, in fact, appropriate.

Does the process envisaged slip between the horns of the original dilemma? It is plainly an attempt to fashion a "skeptical education," and one might worry that the skeptical voices are too powerful to enable the citizen body to make a seriously informed decision, or that the role given to skeptical voices is too weak to inspire broad confidence in the fairness of the decision. Those worries can't be answered from the philosophical desk chair. They should be addressed by running experiments, seeing whether groups of citizens who engaged in this type of process could reach what they saw as responsible and informed evaluations and whether their assessments could gain trust among the wider public. Sophisticated political theorists—like Fishkin—are obviously in the best position to design and carry out experiments of this sort. They may also be able to refine—or radically revise—the structure I have proposed. My aim has not been to arrive at a definitive solution to the problem of political entanglement, but to pose a clear dilemma and indicate ways in which we might respond to it.

I close with two remarks. The first concerns the relationship between the philosophical proposals of this book and the political research carried out by those interested in realizing deliberative democracy. Deliberative polling, as well as the citizen juries of the Jefferson Project, endeavors to find structures for improving democratic decision making. My attempt to understand the relations between scientific expertise and democratic values identifies places at which those structures might be put to use. If you like, Fishkin's question is "How?" and mine is "Where?" The extension of his approach to the domains in which (as I've argued) it is needed requires the modification of some aspects of deliberative polling that, as they stand, fit easily with his preferred uses. I have offered some preliminary suggestions about how those modifications might go (the three-stage process just outlined), but, as already acknowledged, appropriate adjustments should be determined empirically.

My second remark addresses the concern that the situation of our democratic discussion is currently so dire that no redress along the lines proposed is possible: there will always be loud voices decrying any efforts to rebuild trust in expertise. Whether or not I am right about the specific places where Science—or, more generally, public knowledge—and democratic values do not fit, I hope this book offers a persuasive analysis of a *predicament*. All of us should be aware of the limits of our individual knowledge, and of the need to draw on the better opinions of others, if the personal and collective projects we value most are to have any chance of success. If many of us are often in a condition of irremediable ignorance (or worse), then my discussions may help advance the widespread recognition of this state. Once that recognition is achieved, the problem of restoring some type of trust will be vivid for all, for people who adamantly oppose one another, for those who are most alienated from the officially favored "experts." In a situation where a large majority can appreciate that predicament, there will be pressure to find conduits between Science and the people—and perhaps to give the representative bodies, whether in the form of deliberative pollsters, citizen juries, or something entirely different, a serious chance.

Chapter 9

ACTUAL CHOICES

38. THE HISTORY OF LIFE

This final chapter will return from the more theoretical ideas explored in its predecessors to some of the examples with which we began. The hope will be to show how the proposed framework bears on decisions democratic nations now face. Of course, given the perspective on values and democracy advanced earlier (chapters 2 and 3), the conclusions drawn here can only figure as preliminary proposals—the ultimate authority resides in ideal deliberations, and those might diverge from the lines of thought I offer.

Four examples will probably suffice to illustrate the theory of the integration of Science with democracy developed above. The instances chosen—the history of life, biomedical technology, genetically modified organisms (particularly foods), and climate change—will be examined in that order. The sequence reflects my judgment about the relative importance of the issues: it proceeds from the least consequential to what may well be the most significant political question of our age.

Probably more attention has been devoted to the strident attacks on Darwin's evolutionary theory than to any other public scientific controversy. Many outstanding scientists, including Stephen Jay Gould (1977), Richard Dawkins (2009), Niles Eldredge (1982), Francisco Ayala (2007), Kenneth Miller (1999), and Jerry Coyne (2009), have devoted significant effort to combating first "Creation Science" and more recently "Intelligent Design." The hours of neuron use thus consumed have been lost to more worthwhile pursuits. Other writers, including historians and philosophers, have contributed their time: I have written two books on creationism and intelligent design (Kitcher 1982; 2007), and my own efforts have been outstripped by those of others; for example, the distinguished historian Ronald Numbers

(1992) and the tireless philosopher of biology Michael Ruse (1982; 2005). Nevertheless, the controversy persists. Politicians continue to declare that "evolution is just a theory," and in some parts of the world (the United Kingdom, Islamic countries) the opposition to Darwin is growing.

Properly understood, the debate is not merely about evolution or about Darwin. Most of the dissenters want to make room for a position incompatible with large parts of many sciences, including cosmology and the physics of radioactive decay. Darwin serves as a symbol—a Great Man to be defended or a useful punching bag. Symbolic significance is central to the dispute. *For, in the end, it is hardly one of the world's greatest tragedies if the schoolchildren of a particular nation, or of the world, are presented with alternative "theories" of the history of life.* "Equal time" in the biology classroom is a bad thing, for it requires qualified teachers to pretend that brilliantly successful pieces of Science are on a par with contrived bits of religiously inspired doctrine that devout, but honest, scholars abandoned one or two centuries ago. Lying to the young is not a healthy practice, and it may have the effect of distorting their beliefs. Yet the other examples to be considered have more direct consequences for human welfare. As we shall see, the failure to resolve disputes about biomedical technology, about the safety of genetically modified foods, and about climate change, is likely to cause vast amounts of remediable human suffering and death. Holding false beliefs about the history of life cannot aspire to any such impact. Why, then, all the fuss?

Simply this: the opposition to "evilyoushun" is an affront to the proper authority of Science. Darwin's symbolic value for his champions lies in his being credited with one of the greatest advances in the history of the sciences, one for which the evidence is as firm and as broad as that for other discoveries that nobody—not even a pandering politician—would dismiss as "only a theory." Darwin stands for Science—or, more exactly, for a proper division of epistemic labor, one that is needed if sound policies are to be crafted. No appeal to value-judgments needs to be made (or so it is supposed), but rather a sober exposition of the objective evidence. If that can be done, and the authority of Science re-established, we can hope for saner political discussions across the board.

In fact, despite the impressive evidence that scientists (and their philosophical allies) discern in this case, the history of life constitutes a problematic battlefield on which to campaign for the overriding authority of the sciences. For opposition is driven by something particularly deep and hard to eradicate; to wit, a form of chimeric epistemology that protects against the evidence so lucidly expounded (see chapter 6). The religious assumptions behind that chimeric epistemology are thoroughly false, but they will not be abandoned without considerable effort on a wide range of fronts. Some of Darwin's most dedicated and brilliant champions, most notably Dawkins, attempt to uproot the assumptions by direct assault. Not only is this campaign likely to fail, but it appears to be counterproductive. In my youth, opposition to Darwin within Britain was confined to a small group of cranks, figures of fun who understood neither the history of life nor the Christianity they professed. Recent polls suggest that citizens of the United Kingdom who accept Darwinism are now in the minority. Apparently, if you explain to people that they cannot have religion and evolution too, God's stock tends to rise.

I have argued (§27; Kitcher 2007, chap. 5; 2011b) that the *practical* task of creating the conditions for fully secular public reason is extremely complicated, requiring the development of social and intellectual resources to replace what religions have traditionally supplied. Until all that work is done, challenges overtly aimed at Darwin and really opposing the orthodox picture of the history of life (to which Darwin made fundamental contributions) are likely to recur. It seems improbable that this will be the battleground on which the authority of Science achieves its victorious vindication. Scientific (historical, philosophical) effort might better be directed toward other disputes in which challenges to scientific claims directly threaten human welfare and where religious entanglements are less conspicuous. In the interim, children can be protected against games of make-believe in biology class by abandoning the policy of gallant attempts to persuade the general public and leaving the hard work to dedicated institutes that monitor and respond to the local efforts of creationists and "intelligent design-ers." (In the United States, the National Committee on Science Education has done yeoman service in this regard. Under the inspired direction of Eugenie Scott, it has protected the rights of schoolchildren to be told the truth.)

Genuine democracy would be advanced, and the campaigns of Creationists blocked, by applying some of the ideas about dissent offered in the previous section. If it became widely understood that free inquiry is an ideal to be achieved not simply by letting people shout at one another in some public space but by regulating the arena to promote fruitful discussion, we might avoid the needless repetition of answers to the same stale challenges. For over thirty years, in a variety of fora, Darwin's attackers have brought forward a number of charges. Not only have they said the same things again and again and again (and then again), but the issues they raised were once seriously discussed within the scientific community and were carefully resolved long ago. Scientists who have responded to the attacks are thoroughly familiar with the phenomenon. A charge is made—for example, that evolution under natural selection is incompatible with the second law of thermodynamics—and it is decisively refuted. Perhaps this occurs in a public debate. The Creationist spokesman moves on to a new audience, before whom he trots out *exactly the same patter*. No attempt is made to take the discussion further by arguing against the refutation. It is simply ignored—for, after all, the new audience will not have heard it. (The prominent Creationist, Duane Gish, was a master of this strategy of unmodified repetition, but it is shared by his comrades in the "Creation Science" and "Intelligent Design" movements.) If ever there was a form of dissent crying out for a body to certify that nothing new was being offered, it is the succession of challenges offered by the opponents of Darwin.

After several major court cases during the past decades, the time has surely finally come for some licensing of further discussion of these questions. The debasement of discussion through the tiresome reiteration of the same points is a familiar phenomenon: most of us have sat through deliberations in which somebody is answered many times and yet still persists, unaffected by any of the responses. If the trials (in Arkansas, Louisiana, and Pennsylvania) are recognized as having offered fair opportunities for the expression of dissent, it might be appropriate to declare *in the name of healthy free speech* that those who engage in further dissent on this issue should first have to convince some impartial body that they have something new to say—and, if they cannot, that their public utterances have to be accompanied with clear announcement of their failure.

39. BIOMEDICAL TECHNOLOGY

The spectacular advances of molecular biology during recent decades have revolutionized our understanding of the mechanisms of life at a pace whose only precedent in the history of the sciences is the explosive development of physics in the early modern period. Nor is the molecular revolution over. It is reasonable to expect, in coming decades, further advances in understanding intracellular metabolism and the development of multicellular organisms. So there arise questions about what to do with the knowledge so far achieved, about what further lines of investigation to pursue, and about what tools are permissibly employed in both these endeavors. The last topic, especially, raises important choices that have given rise to much controversy.

From the perspective of well-ordered science, all three types of questions are to be settled in the same way; namely, through the best replication we can manage of a discussion among tutored representatives, under conditions of mutual engagement, of the range of human perspectives. In advance of instituting bodies that might simulate an ideal conversation, what options can be identified as particularly salient for them, and what might we expect them to decide?

The autonomists of §19 fear that the ideal of democracy to which well-ordered science is committed would slight "basic research." We saw earlier how such research might be defended in two different ways, by appealing to the value of satisfying human curiosity or by indicating the long-term benefits of addressing "fundamental" questions. When molecular knowledge is considered in the context of medical possibilities, potential diagnoses, treatments and cures, emphasizing the satisfaction of curiosity can easily appear dilettantish—especially when the curiosity is typically felt by the investigator and a small group of colleagues. Yet the career of molecular biology, from its distant roots in classical genetics to the multibranched enterprise of the present, is a splendid advertisement for the strategy of assembling resources through patient attention to the details of structures and processes that can be examined most easily in particular organisms. Morgan's decision to study patterns of heredity by scrutinizing populations of fruit flies; the identification of DNA as the hereditary material (by Avery, MacLeod, and McCarty, and Hershey and Chase, working with bacteria); the understanding of the roles of RNA through experiments on bacteria; the recognition of mechanisms of gene expression in bacteria and bacteriophage; the mapping of a bacterial chromosome; and the subsequent development of sequencing techniques—all these important preliminaries to contemporary molecular medicine, as well as a host more, underscore the value of not tackling complex human diseases directly. The road to success is paved by clever proposals to study some important molecule or process in an especially tractable organism. Ideal deliberation of how to go on should be informed by history, and the history trumpets the importance *for practical projects* of "basic research."

Well-ordered science would not take the form autonomists fear. Rather than eliminating "basic research" in molecular biology, it might well ask for more of it and even absolve aspiring researchers from having to claim that the investigations they propose will deliver medical benefits on an absurdly compressed schedule. The Human Genome Project should have been attractive to ideal deliberators, even without the dubious promises of immediate cures.

Yet well-ordered science would also be alert to opportunities for relieving human suffering. As §19 suggested, it might conceive those opportunities more broadly than the contemporary research agenda does. Under the aegis of the fair-share principle, it would seek ways in which current knowledge could be applied to address diseases that kill and disable a large number of people, especially children, in the nonaffluent world. Part of its plan for research might well include a major vaccine program, based on using molecular tools and designed to supplement the vaccines that work well in rich societies with others that can be exported to parts of the world where conditions are more difficult. It would almost certainly be uninterested in disease research focused on the minor troubles of the well-to-do.

The largest controversies in contemporary biomedical research swirl about proposals to use particular organisms (or organic entities) as tools in investigation. Should investigators be able to create blastocysts, enabling them to derive stem cell lineages, which can then be employed in projects aimed at developing tools for combating neurodegenerative diseases? Is it permissible to clone human beings?[1] What constraints apply to the use of nonhuman animals in experimental inquiries? The framework developed in previous chapters views such questions as properly answered in the ideal deliberations well-ordered science seeks to replicate. What answers might we expect?

Start with the case of mammalian cloning. To clone a mammal, you take the nucleus of a cell from the mammal you want to clone, insert it into an enucleated ovum from a mammal of the same species, stimulate it so that it starts dividing, and insert a small cluster of derived cells into the uterus of a mammal of the same species. A woman who wanted to give birth to a clone of some particularly admired individual could arrange for the nuclear DNA of that individual to be inserted into one of her own enucleated eggs, for the result to be stimulated so an early-stage embryo formed, and for the embryo to be implanted in her womb. If she wanted to "reproduce" some public figure, Tiger Woods or Sarah Palin, say, and if she could obtain the appropriate DNA, she could try to satisfy her (curious) desire. In fact, she would probably be well-advised to make the attempt on an extended scale, since mammalian cloning still requires a large number of embryos to yield a healthy birth.

Well-ordered science would frown on this practice for several reasons. First, the tutored conversationalists would realize how misguided the whole notion of "reproducing" another person is. The attributes of the adult Tiger Woods or Sarah Palin have been produced by a long and complex process in which the nuclear DNA has played some role, but those traits have also been affected, possibly profoundly affected, by a sequence of environments, beginning with the *cytoplasmic* environment in which the nuclear DNA originally found itself, followed by the *uterine* environment in which the embryo grew, and, most evidently, the many physical and social pressures that have impinged since baby Tiger and baby Sarah emerged. None of these environmental influences can be replicated. Perhaps the misguided mother-to-be can hope for an offspring somewhat like a twin of the figure she so admires, but a twin so generationally separated that the resemblance with the model would probably be remote.

Second, insofar as discussants are committed to the conception of the good commended in §7, according to which serious chances for a worthwhile life are central, they will object to the venture of trying to create a child in any specific human image. Choosing one's own life project is a central feature of a good life, and when a parent seeks to impose a particular structure on a child's existence, as James Mill notoriously did with respect to his eldest son, that choice is threatened. The would-be mother interferes with her child's opportunity to choose an individual pattern by her attempt to replicate someone else.

Some instances of cloning might evade this objection. Imagine two women who love one another deeply and wish for a child biologically related to both. If one provides the enucleated egg and the womb, and the other the nuclear DNA, they obtain what they want—without trying to prescribe a particular image toward which their child should develop. This is surely more benign, but it is vulnerable to a third concern.

Apologists for cloning, or for other techniques of assisted reproduction, sometimes defend their recommendations by declaring that they are expanding people's reproductive choices. The defense has a persuasive air about it, because of the suggestion of increased freedom. As usual with abstract appeals to freedom, it is worth asking about the distribution of freedom: exactly whose freedom is being promoted (§11)? Well-ordered science would note first that, as things stand, human reproductive cloning would demand a massive investment in harvesting and modifying ova (because a couple of hundred are required to achieve a single success), would consequently be expensive, and would be confined to the most affluent members of the most affluent societies. More fundamentally, efforts at assisted reproduction divert medical resources from the serious problems of the many, and that should not be done lightly. Perhaps the ideal conversationalists might decide that the loss of the ability to produce a biological child is sufficiently important to sufficiently many people to warrant continued research on and applications of the most successful and promising techniques—although they might also conclude that the world would be better if people were encouraged to think of adopted children as "their own." It seems overwhelmingly likely that they would view human cloning as a bizarrely complex way of achieving goals of relatively minor significance in comparison with the urgent health needs of the world's poor.

Turn now to the decision about creating blastocysts to generate stem cell lineages for research. Here, as with the example of the history of life considered in the previous section, the verdict of well-ordered science is easily predicted. A human blastocyst is a hollow sphere of cells at a full developmental stage before the process, gastrulation, through which the prepattern of the central nervous system is laid down. Human blastocysts are embryonic human organisms—that is, they belong to the species *Homo sapiens*. Because they do not yet have the prepattern of the central nervous system, they lack any neurons and are thus incapable of feeling. What might stand in the way of using them as research tools? The obvious answer: these "clusters of cells" are human beings in a richer sense; they are equipped with immortal souls. Nothing like that answer can, however, be admitted into the ideal conversation. The discussants will know the orthodox scientific story about how human blastocysts come to be: they will understand the set of reactions involved in the taking up of DNA by the fertilized egg and the processes that occur in early cell division. Nothing in this story does anything to allay puzzlements about the "ensoulment" of this tiny organism. Nor can religious sources, certified as almost certainly false, be allowed into the deliberation. Finally, the sufferings of people who might be helped through stem cell research will be fully appreciated, and lines of inquiry promising to alleviate that suffering will be consequently favored. Well-ordered science would tear down the barriers that currently block valuable investigations.

Yet, as with "evilyoushun," it might appear that any attempt to simulate the ideal conversation in societies in which chimeric epistemologies are prevalent is doomed. If the group of citizens recruited to serve as a facsimile of the ideal conversationalists is sufficiently diverse to be convincing, it will contain people for whom the testimony of the scriptures (as interpreted by their religious leaders) trumps everything else. The cognitive conditions on the discussion will be flouted, and that will be expressed in a failure to reach consensus. That is an understandable worry, but I think it is worth trying the experiment.

For there are three sorts of well-established information that might give even deeply devout people pause before they invoked their conversation-stopping veto. Typically, none of these figures vividly in the minds of actual opponents of proposals to create blastocysts. First, the medical condition of those who suffer the degenerative diseases for which cures and treatments are sought needs to be carefully explained, both to create genuine sympathy for the afflicted and for those who love and care for them, and to make clear the considerations that move researchers to think that exploring stem cell lineages might open up new opportunities (that should be done soberly, without hype and premature promises). Second, the mechanisms of early human ontogeny should be described in enough detail to raise puzzles about the exact stage at which "ensoulment" might occur. Third, a part of public knowledge from outside the natural sciences is relevant here: the history of learned disputes among theologians about the "ensoulment" of the fetus. That history would not only reveal how easily the supposedly crucial parts of the scriptures can be interpreted in alternative ways but would also shake confidence in *any* of the rival arguments. Perhaps those initially inclined to think of an absolute decree that could not be violated would be moved by the combination of sober molecular description of ontogenesis and tangled theological history to worry that the metaphysical picture they had taken for granted was dubious, and, moved by the sufferings caused to patients (grown people with sensations, thoughts, and feelings), decide that the prospects for the afflicted could not be held hostage to metaphysical speculation. As I have said, the hope may be unwarranted, but the experiment seems worth trying.

Ideal deliberation about the use of nonhuman animals in research might yield *more* restrictions rather than fewer. Discussants would proceed from an understanding of the goals of a line of inquiry and of the animal suffering entailed in pursuing it. Quite possibly, they would find some research unwarranted because the knowledge sought was too trivial to justify the pains inflicted, or because the chances of success were too low, or because the research could easily be advanced by using organisms with a more rudimentary power of sensation. They might insist on interventions to manage animal pain (as inquiry produces better palliative drugs, it will be easier to use animal subjects without making them suffer). Finally, they might *extend* the possibilities of experimentation by permitting some human beings to volunteer for procedures, when they clearly understand the consequences and freely choose to elaborate their life projects by doing so. Especially noble people, knowing themselves to be dying, might clear-headedly commit themselves to serve the cause of alleviating future suffering. It is interesting to imagine a lifelong lover of animals deciding to volunteer as an experimental subject in research designed to minimize the pains of nonhuman animals—and thereby self-consciously giving expression to the idea of a contract between ourselves and our nonhuman relatives that makes demands on both parties.

Our actual treatment of experimental animals could probably be greatly improved by attempts to replicate the ideal discussion. Those attempts would have to be informed by what is known about the neural bases of pain and the ways in which neural structures are shared across groups of organisms. They would also have to be very clear about the proposed benefits allegedly gained from the research planned, the chances of attaining them, and the possibilities for inflicting less suffering or using animals with less sensitivity. Interestingly, the historical achievements from which so many of our contemporary molecular tools descend are largely based on organisms about whose treatment it would be very strange to worry. Even the most passionate champion of the ethical rights of animals is unlikely to shed many tears over what has been done to *E. coli*.

Well-ordered biomedical practice would, I suggest, retain an emphasis on "basic research," but it would differ from the status quo in at least three ways: it would be far more concerned with the medical needs of the many people whose lives are warped or truncated by diseases that do not afflict the affluent (and it would be correspondingly intolerant of research that responds to minor discomforts of the rich, including lines of investigation that "enhance the opportunities for assisted reproduction"), it would open the way for using clusters of human cells (early-stage embryos) as tools in research with serious chances of achieving medical benefit and would not permit religious objections to block such research, and it would be more sensitive to issues about the use of experimental subjects, human and nonhuman. These three apparently disparate modifications are bound together by a common theme, one unsurprising in the medical context: the strong desire to understand and minimize pain and suffering. Bacteria and blastocysts are not subjects of pain and suffering; poor people ravaged by infectious diseases, victims of neurodegeneration, and maltreated laboratory animals all are. Religions that inspire defective chimeric epistemologies arbitrarily stop the conversation and interfere with the relief of suffering. Yet, when these religions repudiate their claims about the "sacredness of the embryo," it is worth recognizing the value of their emphasis on "sacredness" (Dworkin 1993)—although a substitute word with similar resonances might be helpful. Living organisms, especially sentient living organisms, animals capable of feeling pain, should not be described in the language of a mechanistic biology, as if they were systems for us to handle as we will. The proper legacy of the religious emphasis on the "sacred" is a type of humanity that secular thought should work to preserve. Well-ordered biomedicine is thoroughly sensitive to suffering, wherever it arises, and dedicated to alleviating it insofar as is possible. In that, it is thoroughly humane.

40. GENETICALLY MODIFIED ORGANISMS

The villains of the previous sections, the people whose views currently block movement in the direction of well-ordered science, are disproportionately found in the United States. Controversy about genetically modified foods (GMFs), and genetically modified organisms (GMOs) in general, however, rages most fiercely in Europe. Opponents of GMOs press two different concerns: first, releasing these organisms into the environment, whether as crops or as means of protecting crops (bacteria can be employed to prevent ice buildup on fruit trees or to repel potential pests), would risk irreversible damage to the environment; second, the socioeconomic consequences of introducing the organisms would be catastrophic, for they would displace traditional forms of agriculture and reduce vast numbers of poor people to a condition of economic dependency verging on slavery. The arguments are connected. Because the critics view the large agricultural companies as hell-bent on making exorbitant profits, they dismiss scientific reports to the effect that the environmental risks are minimal as reflecting monied interests—in effect, the research has been bought. As in the examples of the two previous sections, the arguments are advanced in a public arena in which ignorance abounds. Unlike their American counterparts, European high school students are often inclined to mark as true the statement "GMOs contain genes, but ordinary organisms do not" (Jasanoff 2005). People who grow up with such beliefs easily accept the picture of genes as mysterious little agents of evil, inserted into healthy foods by the wicked minions of agribusiness and destined to leak out into the environment with devastating results.

The introduction of *any* new organism into the environment brings risks with it. The history of agriculture and of conservation efforts has often brought unpleasant surprises. Breeding rabbits in Australia caused the extinction of indigenous marsupial species. As a desperate remedy, ecologists tried to control the damage by infecting rabbit populations with the *Myxoma* virus. The virus spread rapidly, eventually making its way across the world, and the Sussex hills of my childhood were dotted with putrefying rabbit corpses. Further cascades of consequences affected predator and plant populations. The long-term impacts of pesticides, most famously DDT, are also well known. There is nothing special, or especially risky, about *genetic* modification of organisms.

If agriculture is to progress, it will be through producing crops that are more fruitful or that are able to grow under a wider range of conditions, and that can be done either by modifying the crops or by finding ways to enable existing crops to thrive in habitats they cannot presently tolerate. You can try in advance to assess the impact of the new strains or the modified environments, as responsible agricultural practice has done for decades if not for centuries, but however much testing you have done, there will still be questions about what will occur once the novel techniques are applied broadly and the corresponding risks that damage will ensue. *Quietism* in agriculture is the position that we have achieved enough. Our ancestors have bequeathed to us a basket of crops and methods of environmental modification, and we can now employ these to sustain the world's population. They have staggered their way through a sequence of experiments—and some disasters—but we can now be grateful that the process can stop. No further risk taking is needed.

A moment's reflection shows that quietism is untenable. Its claim to have eliminated all risk is spurious, for we cannot be completely sure what will happen if current agricultural practices are sustained indefinitely. Nobody knows for sure if the continuing practice of growing our food in the ways we do will accumulate substances in the soil that eventually make further harvesting of those crops impossible. Nobody knows what changes in the earth's climate (whatever form they take) will produce and whether current patterns of agriculture will be able to adapt to them. Yet besides these abstract points, there is a more concrete rejoinder to quietism. For many of the world's people, particularly in Africa and parts of Asia, current agriculture is unable to provide them, in the environments in which they live, with ways of reliably growing the food they need. Some people inhabit regions affected by periods of extreme drought, others experience temperature fluctuations, or sudden incursions of pests, or soaking rains. Opposition to GMOs is largely a European phenomenon, not much heard in the land of the potential producers (North America), nor in those of the potential consumers (Africa, Asia). In fact, many of the spokesmen for the world's poor are impatient with what they see as the scruples of people who do not feel any threat of starvation (Stewart 2004, 161).

Well-ordered science would respond to the agricultural needs of the poor as it does to the neglect of their health. An ideal deliberation would view the elaboration of Science to improve crops and crop yields as an important direction for research, especially in the form of producing crops able to grow in conditions where poor, malnourished people now live. GMFs and GMOs are potentially valuable tools in this effort. Realizing their promise requires research to gauge the effects of introducing them into the kinds of environments where they are needed. Once the quietist dream of a risk-free agricultural practice has been abandoned, it becomes evident that the appropriate standards are exactly those used in any responsible agricultural innovation. Investigators are charged with envisaging the various routes along which environmental damage could occur and exploring whether the organisms under consideration would actually pose these types of threats. (For an excellent review of the character of these investigations, the misconceptions of them that have dominated public discussion, and the possibilities for making surveillance even more vigorous, see Stewart 2004). Although it is sometimes suggested that GMOs are particularly worrying because the "alien genes" they contain could be transferred to other organisms with unpredictable (and horrifying) results, this fear rests on an error well-tutored discussants would recognize. Any novel organism that breeds true (as crops and organic agricultural tools must) will have a genotype that varies from the standard stock, and the genomic novelties it harbors, whatever they are, have the potential to spread through hybridization.

The studies well-ordered science would recommend would evaluate the risks according to *recognizably* high standards of agricultural experimentation and trial. Transparency is crucial here (§24). For that reason, release of GMOs would have to be preceded by field studies carried out under the aegis of some clearly independent public body, supervised by the groups of diverse citizens that play so crucial a mediating role in well-ordered science. Assessments of safety must not only be free of the taint of business interests, but must also be clearly perceptible as beyond profit's corrupting influence. Given this precondition, the discussants would reject any blanket ban on GMOs, encouraging research to deploy molecular tools and to probe thoroughly the environmental safety of the products. They would license GMOs whose harmlessness was certified by high standards, and would hope for agricultural innovations of this sort to improve the nourishment of the poor.

So far, I have attended to only one strand in the opposition to GMOs, the concern about risk. Addressing the concern is straightforward in well-ordered science, and it would not be hard to approximate the ideal in practice. Once it is understood that quietism is untenable, once it is recognized that GMOs and GMFs pose no special risk (they are not distinguished from other organisms by containing special evil ingredients), once it is apparent that responsible agriculture has procedures for evaluating risk, and once there is a trusted public agency for assessing risks independently of the influence of profit-seeking corporations, the stage is set for the piecemeal production and licensing of GMOs. As with medical research, we might also hope that broader recognition of the plight of the poor and malnourished would generate a sympathetic movement to direct research toward innovations suited to the environments in which the unluckiest people live.

Recall, however, the second theme sounded by the critics. GMOs will be used by agribusiness to create a form of economic independence, verging on slavery. The underlying argument begins with a banal point: successful agricultural innovations displace older practices. By itself, that is unproblematic: we typically seek new ways of doing things because current methods are unsatisfactory, and, faced with the dire needs of the malnourished, it would be sentimental nostalgia of the least defensible sort to insist on retaining "traditional practices" when new techniques allow the children to be fed. The problem arises because the suppliers of the new crops can design them so that a local farmer must keep returning, season after season, to buy a fresh supply.[2] GMFs are likely to contain two sorts of modification, one that fits them to overcome some environmental challenge (they are drought-resistant, say) and another that prevents them from setting seed. Even before genetics achieved the current level of molecular understanding, experimental agriculture was sometimes able to do this in producing improved crops (hybrid corn, hybrid cucumbers), but the enhanced abilities for manipulation of the genome permit a far more delicate, effective, and systematic practice. The price of acquiring the new crops, able to withstand periods of drought (for example), is that the farmer cannot harvest seed at the end of the season. Each year, he must pay the agricultural company again, and, because the alternative is a serious risk of starvation (especially for his children), he must pay what the corporation demands.

The core of the serious opposition to GMOs and GMFs is a vision of how the agribusiness market can be expected to operate. If beneficial crops are developed and introduced, with acceptable risk, so as to provide food for the poor, it will be subject to a condition in which profit maximization will drive the prices of supplies to a level just below that at which desperate people would prefer to risk starvation. The poor will remain in a state of penury, while a few executives in the developed world draw colossal salaries. Whether or not capitalist agriculture is quite as rapacious as the critics portray it, the asymmetries are evidently disturbing. Even if the executives are kind and forego applying the screws, the availability of basic resources like food should not be subject to noblesse oblige. The problem is that knowledge needed for the public good has become privately owned.

Well-ordered science treats knowledge as public property. Consonant with the approach to values it presupposes, arrangements in which the preconditions of a worthwhile life for some people (in this instance, many people) are controlled by the wishes of a few cannot be permitted. One way to protect people in a situation where something necessary for their existence is privately owned is to regulate the use that can be made of that private property. Another, simpler solution is not to allow private ownership in such instances. That has seemed to many innovators exactly the right attitude to take to the products of their ingenuity and labor. According to a famous (apocryphal?) anecdote, when asked if he intended to patent his polio vaccine, Jonas Salk reacted with surprise at the idea: "You wouldn't patent the sun," he replied.

Ideal discussants might well adopt Salk's attitude. Yet their conversation, given the variety of perspectives it would have to include, would surely consider the suggestion that private ownership of knowledge (intellectual property) is a valuable institution for generating rapid advances. The suggestion deserves their (and our) scrutiny. For it is not that individual scientists are typically thrilled by envisaged economic returns. If market forces spur their steps toward the lab early in the morning, the market is one that operates on rather different principles, that holds out the lure of credit (they want, in Mayr's phrase, "to be right"—§33). The point must be that the support of corporations can provide facilities for scientific research, expanding the range of investigations beyond what the public purse could fund. Ideal discussants would have to consider whether the likely expansion of successful investigations would be worth the loss of direct control of the knowledge achieved. Yet any engagement with the predicament of the poor would surely prevent them from abdicating control completely. Even if they did not conclude by endorsing the Salk solution—all knowledge should be publicly administered—attention to the valid point at the core of the opposition to GMOs, the danger of corporate extortion, would lead them to impose serious regulation on the prices and profits to be permitted.

We live in a world with a ramshackle system of intellectual property, one that has developed in the same haphazard way as the institutions of Science as a whole. It urgently needs rethinking. A first clear step toward well-ordered science would be to impose constraints on the uses corporations can make of what they think of as "their" property, when what they "own" is urgently needed by the poor. Renewing the ethical project (chapter 2) requires mutual engagement with all, and that cannot be done without recognizing the obligation of ameliorating the condition of needy and malnourished people (especially children). Limiting the profit seeking of agribusiness is a corollary of that, one that would liberate molecular tools for the greater human good.

41. CLIMATE CHANGE

For thirty years now, we have been warned. The emission of greenhouse gases into the atmosphere, even if we take immediate steps to check it, will raise the mean temperature on the earth's surface at least 2ºC by the end of the century. Although the consensus among climate scientists on anthropogenic global warming is overwhelming, the reality of the effect continues to be denied by a few fringe scientists—distinguished physicists and ex-physicists and meteorologists among them—whose articles and speeches influence significant portions of the public, particularly in the United States but also elsewhere. Climate deniers have successfully blocked attempts to introduce policies for coping with potential environmental change, and they have capitalized on recent events to suggest that the alleged consensus among climate scientists stems from an antibusiness ideology. They have exulted over a "mistake" made by the Intergovernmental Panel on Climate Change (IPCC) in announcing the imminent demise of the Himalayan glaciers and have delightedly quoted from the "conspiratorial e-mails" among members of the East Anglian climate center. In the wake of these assaults on the consensus, and the roughly simultaneous failure of the Copenhagen climate summit to craft any agreement on international strategy, polls reveal that public concern about climate change is fading, even in the more environmentally conscious countries.[3]

This is an enormous tragedy, a huge failure of worldwide democracy. Most people, including most of those who oppose, or are indifferent to, any policies for addressing the problems that will be generated for our descendants by our continued excessive use of fossil fuels, care deeply about the opportunities their descendants will have. Nor are such concerns narrowly parochial, focused on *their* grandchildren and great-grandchildren. When natural disasters strike faraway people, television reports of the plight of the survivors prompt many to open their hearts and their checkbooks. The failure to treat issues of climate change with the seriousness they deserve stems from many forms of ignorance and many causes.

One important factor, as we shall shortly see, is the genuine difficulty of some of the issues. Yet public discussion is muddied, from the outset, by confusion about the most basic feature of our situation. *Level 1* of the debate about climate change concerns the truth of the statement at the heart of the consensus among climate scientists: *even if steps are taken now to limit the worldwide emission of greenhouse gases, there will be a mean rise of at least 2°C by century's end.* Many climate scientists would argue that this minimal claim underestimates the dangers, but the fact that it remains in doubt is an enormous handicap to pursuing the debate at all its levels. At level 1, discussion should already be over, and the public—the entire human population—should be inspired to take up the more complex questions that emerge at higher levels.

Our inability to reach population-wide consensus on that first level results from many of the problems previous chapters have considered. Citizens of affluent nations are confused because there is an apparent debate among conflicting "experts" (§30). They do not appreciate the fact that the "experts" are not specialists in the fields on which climatic predictions depend. Nor do they understand the tight connections linking many of the deniers to special-interest groups (particularly oil companies; Oreskes and Conway 2010). As they think of the future, they worry that the "antibusiness lobby" favoring environmental action is likely to diminish the prosperity of their grandchildren, as well as disrupting their own lifestyles. They are especially vulnerable to the alleged "scandals" that have been exposed in the climate science community. For they are in no position to assess the evidence on which the consensus claim rests, nor to appreciate that the "mistake" on the part of the IPCC occurred in a single sentence in a long draft, a sentence attributing an overly rapid rate to the melting of the Himalayan glaciers—*and that there is a large amount of independent evidence for the fact that those glaciers are melting, albeit at a lower rate*. When they learn that climate scientists wrote some negative things about their competitors in unguarded e-mails, they have no basis for setting that in the context of the complex competitive-cooperative interactions that pervade contemporary Science (§§33–35); nor when they learn that these e-mails mention "tricks" do they understand that this is not a reference to illicit chicanery but to graphical methods of presenting a conclusion perspicuously. They sense that the scientists are making value-judgments, and because they have been taught that Science ought to be a value-free zone, they are suspicious. Their suspicions are deepened when the most effective champions of the consensus view seem clearly committed to values—even if those are values they share. When James Hansen urges action to prevent the "storms" that will threaten the lives of his (and all our) grandchildren, he can easily be portrayed as overstepping the bounds of the scientist's proper role (Hansen 2009; Keller forthcoming).

Well-ordered science would correct all these defects of our public discussion. It would insist on transparency, on laying bare the considerations that motivate the climate deniers. It would replace our imperfect channels of transmitting scientific information, which, in this instance, have failed disastrously: American newspapers have consistently reported the dispute at level 1 as if there were two sides with equal credentials, and, although such organs as the *Wall Street Journal* and the *Washington Times* have been particularly egregious in their coverage, no major newspaper can be proud of its record of enlightening the public (Oreskes and Conway 2010; Leuschner 2011). It would provide a more accurate picture of the internal workings of the sciences, and of the inevitability of value-judgments—thus scotching the idea that the public-spirited scientists who use their deep knowledge to warn others are somehow violating their terms of employment. Under well-ordered science, the level 1 debate would be over, and a concerned human population would be moving on to the more difficult decisions.

There are three further levels at which informed public exchanges are needed, and our discussions should, I believe, be framed by the approach to values, democracy, and the institution of Science outlined in previous chapters. *Level 2* would seek as detailed and precise an account as we can now achieve of the likely consequences of our present practices of fuel consumption. Precision will often be impossible, and it will frequently be necessary to rely on the hunches of informed scientists. Yet steps can be taken to simplify the issues by considering a wide a range of potential effects on human life. Because policies of trying to ward off undesirable outcomes will often entail economic sacrifices, issues about species extinction (notoriously hard to predict, in any event) or other types of environmental conservation are probably best omitted. The chances of enlisting wide public support are increased by focusing on the impact on the future human population.

We urgently need a synoptic vision of the various ways in which global warming will challenge people who live in different parts of the world (something analogous to the index of human needs, envisaged in §20). Increases in mean global temperature will produce rising sea levels, as well as glacier melt. The most evident consequences are inundations of low-lying areas, as well as disruption of supplies of fresh water. Changing weather patterns can also modify the temperatures and the rainfall in regions where large numbers of people live and where they grow their food. Alterations of these types can affect the ability to provide shelter and disrupt existing systems of agriculture. Coupled to these longer-term effects—the disappearance of the Maldives or the Bengal coast, the transformation of Italy into a desert (for example)—are shorter-term fluctuations. If, as many climate scientists maintain, the frequency and violence of the weather will increase, there will be deviations of greater amplitude about a different mean. It is not just that an area currently one meter above sea level will go under water, but that places four meters above sea level will usually be only two meters above and will become flooded in the more intense storms.

Many of these modifications are likely to be conducive to the spread of disease. If the winter snowpack in the mountains is greatly diminished, quick runoff in the spring, even if it does not cause floods, will be followed by periods of drought (California, for example, is heavily dependent on the winter conditions of the Sierra Nevada). Lack of water places stress on the human body, interferes with the production of food, and makes hygiene more difficult: all factors making it easier for disease vectors to spread. Forced human migration is likely to yield overcrowding and food shortages in those areas apparently promising the best opportunities for shelter, again facilitating the transmission of infectious diseases. Flooding of areas where wastes are prevalent will produce polluted waters and advance the same end. Further problems may stem from environmental modifications causing new patterns of interspecies interaction and thus favoring the evolution of new parasites and bacteria from forms that currently only affect other species.

A full catalog of the many possible ways in which the lives of our descendants could be adversely affected—and, of course, of whatever scenarios might make their lives easier—accompanied by the best estimates of probabilities we can give (relatively precise in the case of inundation of low-lying regions of the world, quite uncertain when we consider the evolution of new diseases) would prepare the way for posing a series of fundamental questions. Pick a large number of future people, a billion, say, or two billion. What are the chances of avoiding any course of future events that will wreck the lives of more than this number of people? Even without precise knowledge of the chances of any particular scenario, it may be possible to declare that our inaction on climate change would probably produce disaster for some very large group of our descendants. Ignorance about whom or how or when is compatible with conviction that some form of catastrophe will occur.

Under well-ordered science, the best information we have from a variety of disciplines would be employed to work out this synoptic picture and answer the kinds of questions just posed. Effectively, the level 2 debate closes with a series of specifications of the chances of future populations of various large sizes experiencing disaster (of some unspecified form). *Level 3* proceeds to work out an ethical framework within which policies for apportioning sacrifices can be assessed. Here, too, factual information must be supplied by experts, or at least by the most expert testimony that can be found. The reasonable core of current opposition to climate-change policy is the suspicion that measures to limit the use of fossil fuels would impoverish those who come after us—presumably through the collapse of important industries and some resultant economic meltdown. That suspicion should neither be treated as an indisputable certainty, as some advocates of business-as-usual tend to do, nor should it be entirely neglected. Any reasonable decision about climate change must be based on comparing the anticipated awfulness of the results of inaction with the estimated costs of the various preventive measures we might undertake.

Proposals for limiting fuel emissions are inevitably linked to considerations of alternative sources of energy and of how the world would fare if its entire energy budget were reduced. In this arena, too, there are great uncertainties, and the best we can hope to achieve is to outline a number of strategies for reducing dependence on fossil fuels, to estimate, insofar as we can, their likely effects on human well-being, and so to compare them. The ethical dimension of the discussion lies in considering how the impact of various overall cuts in energy use or of transitions to alternative forms of energy might fairly be distributed. That is best done through the full exchange of perspectives, through a serious attempt to understand the forms of deprivation now felt by people in developing countries, and through an understanding of the problems that would be engendered if the substructure of the nations most responsible for past pollution were suddenly to be undermined. Current discussions about possibilities of cooperative work on climate change are, I believe, hampered partly because of failure to recognize the predicaments of others and partly because of inflexibility with respect to economic arrangements. It is entirely possible that a fully ethical solution to the problems posed for us by climate change will require commitment to a far more egalitarian distribution of the world's resources. In this, the ideal of serious equal opportunities for a worthwhile life (proposed in §7) may be fundamental. *None* of our great-grandchildren may have much of a chance for a worthwhile life unless we come to terms with the fact that economic resources are not life's *summum bonum*, that they are means to more worthy ends, and that distributing them fairly can allow each—and all—of us to live well.

Level 4 of the debate consists simply in discussing—under terms of mutual engagement, it goes without saying—the options presented at levels 2 and 3. How much risk of future suffering are we prepared to allow? How much by way of burdens on ourselves and our descendants are we willing to take on? Given the difficulties of the issues involved at levels 2 and 3, the uncertainties of environmental scenarios and patterns of economic growth, it would be foolish to predict the range of choices that would emerge, and hence the options salient at level 4. Hence, it is impossible to close this section, as its predecessors closed, with an assessment of what well-ordered science might recommend and of how we might try to realize its recommendations in our actual policies. Instead, the ideal of well-ordered science, and the other proposals made throughout this book, offer only a blueprint for a conversation, without any ability to predict how that conversation should turn out.

We urgently need that conversation. As things stand, productive discussion on climate change is blocked by the many serious ways in which the practice of Science in the affluent societies where most research is done falls short of being well ordered. Perhaps the greatest stumbling block is that with which this book began, the erosion of scientific authority. Without a reasonable division of epistemic labor, public debate at level 1—the *easy* level—cannot reach a conclusion, and hence humankind sits, apathetically, allowing our problems to grow worse. Restoring the division of epistemic labor is important, but, as I have repeatedly suggested, other changes are needed too—a deeper understanding of the ethical project and of democracy, a wide-ranging integration of expertise and democratic values. This book has attempted to chart ways in which that integration might proceed. Whether or not the future follows the course outlined here, I hope we achieve the needed integration in the time we have.

NOTES

INTRODUCTION

1. Sometimes, this may involve people who are especially good at concealing what they are doing. Probably far more common are instances in which the oppression is the effect of impersonal agents whose activities are hard for anyone to fathom.

CHAPTER 1

1. I shall follow this convention throughout, using *sciences* to designate fields of inquiry, and *Science* to refer to the institution embracing the diverse investigations researchers undertake.

2. Strictly speaking, contemporary evolutionary psychology divides into a much-publicized scientistic wing, conforming to the "Santa Barbara paradigm," whose best exemplar is the work of Leda Cosmides and John Tooby (for example, Cosmides and Tooby 1992), and a far more modest collection of approaches that make no grand claims about the ability of evolutionary ideas alone to explain human phenomena (Nettle 2011). The scientistic form is the target of my criticism, here and elsewhere—and it is this form that captures the imagination of journalists and political commentators, and thus filters into public consciousness. The modest approach claims far less, takes evolutionary theorizing far more seriously, and is far more likely to deliver some reliable judgments.

3. This should not be confused with what I described as "the division of cognitive labor" (Kitcher 1990). The latter notion (which will occupy us in chapter 8) arises *within* scientific inquiry, in terms of diversification of programs of research. The notion introduced in the text concentrates on separating the roles of scientists from those of members of the broader public.

4. Plato insists on the overall happiness of the city as the ultimate goal, and, in response to the objection that the rulers are going to live demanding lives and be poorly rewarded, he has Socrates respond: "We'll say that it wouldn't be surprising if these people were happiest just as they are, but that, in establishing our city, we aren't aiming to make any one group outstandingly happy but to make the whole city so, as far as possible" (Plato 1992, 95).

5. Of course, Plato does think that, eventually, the *kallipolis* will become corrupted, decaying first into oligarchy, then into democracy, and ultimately into tyranny. This is the story of Books VIII and IX. The first step in decay comes about because the wrong people are given a chance to rule, and this arises because of a failure in discerning the natures of individuals (in effect, Plato's eugenic program goes awry); Book VIII 546a–547a (Plato 1992, 216–17).

6. As Allen Buchanan has pointed out to me, refashioning the system of public knowledge may require reshaping some of the epistemic practices of individuals. Although most of the discussions in the chapters that follow are concerned with the *social* organization of knowledge, some social modifications might entail changes in individual practice: see, for example, §§35–36.

7. My phraseology here recalls a remark made by the critic George Steiner. In 1967, Steiner offered an informal talk in connection with a production of a Molière play that the Christ's College Dramatic Society was presenting at a Cambridge drama festival. He claimed that Molière and Stendhal would always have more to teach us about human thought and character than all the cognitive scientists there ever have been or ever will be. Molière and Stendhal are good examples, but I take Sophocles and Shakespeare to be even more convincing.

8. See, for example, (Dawkins 2006) and (Hitchens 2007). (Dennett 2006) is considerably less combative.

9. As for example, in the case of Green Fluorescent Protein, which my colleague (and friend) Marty Chalfie cleverly managed to insert, and use, in the study of nematodes: work that won him the 2008 Nobel Prize in Chemistry.

10. This is the view of orthodox Bayesianism. For a lucid articulation, and assessment of its merits, see (Earman 1992).

11. Although Larry Laudan is principally concerned with epistemic values (typically elements of probative and cognitive schemes of values), there is some kinship between the ideas I express here and his (1984).

12. Here I am indebted to Gregor Betz.

CHAPTER 2

1. In Allen Buchanan's witty phrase: "things that go 'ought' in the night."

2. This chapter condenses the material of (Kitcher 2011a). Plainly a sketch cannot be as convincing as an extended treatment, one that tries to anticipate and address objections, and I hope that skeptical readers will consult the full-dress account.

3. The thought of an "unseen enforcer" is a very powerful means of increasing compliance with the rules. It has been adopted by an extremely large range of human societies—possibly by all (Westermarck 1924, chap. 50), (Kitcher 2011a, chap. 4).

4. For more details, see §§18–21 of (Kitcher 2011a).

5. Note that allowing for some examples of ethical progress is not to suppose that progress is very common. Nor does it entail that the large historical trend is upward.

6. Even when the possibility of salvaging a notion of ethical truth in the envisaged way is recognized, the notion of progress *from* remains fundamental. It is only because they survive under progressive transitions (understood as modifications that respond to problems) that the statements count as true.

7. I have in mind particularly the !Kung. See (Lee 1979; Shostak 1981).

8. This presupposes that the size of the human population be restricted so that the available material resources suffice. For discussion of this point, see (Kitcher 2011a, §§ 48–50).

9. Ethical discussions fail to be ideal if they *actually* incorporate false assumptions. To appraise them as nonideal, one can sometimes point to an identifiable error. Moreover, it is sometimes possible to judge that the deliberators themselves should have been able to identify that error. I am grateful to Torsten Wilholt for comments that have provoked me to be more explicit about these different modes of evaluation.

10. In chapter 5 of (Kitcher 2007), I argue for the thesis that the substantive doctrines about supernatural beings offered by each of the world's religions are all almost certainly false. The argument will be briefly recapitulated below (§26).

11. Here I draw on ideas present in the eighteenth-century sentimentalist tradition, notably in Smith's theory of moral sentiments. I have explored the mirroring metaphor, as he develops it, in (Kitcher 2005).

12. This is effectively to construct an analog of the "ideal spectator," but one informed by extensive factual knowledge. It articulates further the synthesis of the methodological ideas of Smith and Dewey that I propose in (Kitcher 2005).

13. This does not entail that the entire species is the appropriate group for *every* ethical issue. There are some questions that properly arise more locally (see Kitcher 2011a, chaps. 9 and 10). For issues about scientific knowledge, however, the inclusive conception seems apt.

14. In shaping the conception, and in the formulation given here, my debts to Mill, and secondarily to Dewey, should be apparent.

15. This is prominent in the writings of Richard Dawkins. But the idea is very old—Aristotle's final chapter of the *Nicomachean Ethics* sounds a similar theme.

16. Thus Dewey advances on the Millian formula from *On Liberty*, when he opens *The Public and Its Problems* by posing the issue in terms of the freedom from coercion of *joint* projects.

17. The exceptions are those whose cognitive and emotional possibilities cut them off from fully developed relationships with others. This is, I believe, why we find prenatal genetic testing for some sorts of traits a merciful way of proceeding. These questions are discussed in more detail in the later chapters of (Kitcher 1996).

18. This is because they presuppose claims about the existence of a supernatural being that are almost certainly false. See the last chapter of (Kitcher 2007). There are many other, less obvious ways in which the anti-Darwinian broad scheme of values violates the conditions of ideal discussion.

19. Here I am grateful to Evelyn Fox Keller, who prompted me to be explicit that the serious opposition rests not on "capitalist values" but on a fundamental notion of quality of life. The opponent thinks that imposing restraints on economic growth will send the world economy into a tailspin whose consequences will be wretched lives for all. (Whether this is *correct* is not relevant to the point at hand; to wit, that a contrary case might be founded on serious value-judgments.)

CHAPTER 3

1. There are many valuable contemporary discussions of democracy from different perspectives, and I admire the accounts offered in (Estlund 2008), (Goodin 2003), and (Richardson 2002). My own approach is heavily dependent on ideas of classical authors, particularly Rousseau and Mill, as well as on the insights of the political theorist Robert Dahl. The principal debt, however, is to Dewey.

2. Dewey offers a concise history of the way in which elections came to be viewed as the proper vehicle for public control. He credits James Mill with the "classic formulation of the nature of political democracy" (Dewey 1985, 93).

3. As it stands, this is surely unrealistic, and it may even be a poor ideal if one allows for alternatives deliberately designed to gerrymander in one's own favor ("Instead of the law proposed, I prefer that everybody except me be subject to that provision"). As we shall see, however (§12), the fundamental point about agenda setting is an important one.

4. The standard economic dogma about noncomparability of preferences gained its appeal because of worries that hypotheses of interpersonal utility comparison do not lend themselves to precise operational testing. Despite that fact, we can use psychological and physiological information about responses across our species to make rough judgments. As a pure instance of the scenario given in the text, imagine that two subjects both prefer the same option in a binary choice; the first would be prepared to switch if the cost of staying was to endure a very slight electric shock; the other would not be prepared to switch to avoid any series of shocks, even those that would leave him unconscious. Assuming that neurological and psychological tests reveal no significant response differences in either, I see no basis for insisting we just cannot tell whether the perceived cost of the minor shock for the first is lesser or greater than the perceived cost of the lengthy series of shocks for the second.

5. Some contemporary theorists prefer to approach the notion of negative liberty by identifying freedom as the absence of domination, where *A* is taken to dominate *B* just in case it is possible for *A* to interfere arbitrarily with the actions of *B*. Important variants of this view can be found in (Pettit 1997) and (Richardson 2002). For this—"Republican"—ideal of freedom, the notion of *arbitrary interference* would ultimately be determined by the outcome of an ideal conversation under conditions of mutual engagement.

6. This is Rousseau's important presupposition that not every group can form a genuine social contract. The appropriate groups are subject to size requirements (they must not be too large), and they must satisfy a condition of like-mindedness. See "Social Contract," in Jean-Jacques Rousseau, *The Basic Political Writings* (Indianapolis: Hackett, 1987), pp. 141–227.

7. For elaboration of this idea, and brief defense of it, see (Kitcher 2010).

8. See also (Dewey 1997, 37, 94, 97–98, 100, 260, 359; 1958, 57). It's worth noting that Dewey is specifically concerned to play off Berlin's two concepts against each other, using each to expose the limitations of the other and attempting to combine the insights of both (1997, 91–98).

9. It's no accident, therefore, that when Dewey offers his most extensive treatment of social philosophy, he begins with a generalized version of the question that is central to *On Liberty*, the issue of the control of transactions among individuals (Dewey 1985, 12, 15).

10. It's thus no accident that Mill's political economy is much concerned with questions of distribution, and that his proposals along these lines call for heavy inheritance taxes and the supply of educational opportunities for all. Mill does think that withdrawing excess wealth from the very rich will actually contribute to the development of their children, but, even if he abandoned this psychological hypothesis, he'd still remain committed to the expansion of the opportunities for the poor.

11. This, of course, greatly impressed Tocqueville (*Democracy in America*), although he feared that, with the growth of a more complex society, it would pass. Dewey, who emphasizes self-realization in joint decision making, sees voting and elections as "rudimentary political forms," poor substitutes for discussion and "mutual education" (Dewey 1985, 208, 209)—and aligns his concerns with Tocqueville's analysis.

12. Although Rousseau tries to forge a link between commitment to the common good and voting procedures, claiming that those who are outvoted must recognize themselves as having been mistaken, this cannot hold in general, for the majority might simply be unaware of the dreadful consequences their decision might bring for some members of the population. Indeed, Rousseau's own formulation alludes, vaguely, to conditions that have to be met for votes to serve this educational function. Nor is it reasonable here to appeal to the Condorcet Jury theorem, for what

is at issue is precisely whether or not the individual probability of determining the common good is greater than half for the citizens who form the majority.

13. Many writers would argue that the equal elimination of dominance, interference, and violation of rights for the poor, for women, and for ethnic minorities remains incomplete. For a strong formulation of the feminist case, see (MacKinnon 1989).

14. James Madison, the *Federalist*, no. 10. For illuminating commentary, see (Dahl 1963), especially chapter 1.

15. See (Dahl 1970, 67–68). I amend his analysis by looking at the maximum amount of time per speech, rather than taking that time to be fixed (his choice is ten minutes) and considering the number of issues covered.

CHAPTER 4

1. Vast amounts of philosophical ink have been spilled on the conditions for an individual subject to know something. I follow Edward Craig's important insight that the central role of the notion of knowledge is to mark potential sources of information (Craig 1990). Understanding knowledge in these terms enables one to see that the root concept is that of a *knower*, and that this is bound up with particular role-governing norms. In the "state of nature," the conditions of our Paleolithic ancestors, and of a few contemporary groups, those norms are easily articulated in terms of sincerity and observational responsibility. Observational responsibility, in turn, can be viewed as paying sufficient attention to make it likely that the beliefs you acquire will be true. Thus Craig's insights can be integrated with the reliabilist approach to knowledge, pioneered and lucidly elaborated by Alvin Goldman (Goldman 1986; 1999). Recognizing the Craigian function of the concept of knowledge in social exchange is compatible with appreciating that this concept sometimes has other functions: for example, in the deliberations of individual agents. As Isaac Levi has persuasively argued, the concept of knowledge required to discharge this function can be taken as true full belief (Levi 2010). Reliability is thus relevant for only one function of the concept, albeit the one of central interest in this book.

2. In any society with a complex legal system—and the extensions of codes of law recorded in extant documents make it plain that intricate bodies of law existed before those documents were written—there will be procedures to adjudicate the disputes that inevitably arise. Standards of evidence have to be developed, to assess the reliability of informants. These standards surely served as the starting points for developing procedures for certifying public knowledge.

3. Plato apparently allows a few women into the most privileged group; Aristotle seems to exclude them all.

4. It would be interesting to supplement the brief account I shall offer with

consideration of the systems of public knowledge achieved within Islamic and Jewish communities, and of the interactions among the systems of the three monotheistic religions, especially in medieval Spain. For my purposes, however, Christianity is particularly appropriate, because it produced institutions of public knowledge against which the early champions of modern Science reacted (which is not to deny that those institutions also contributed ideas on which those champions built).

5. A clarification is in order here. The Royal Society was much inspired by Bacon, and Bacon clearly had a plan for the embedding of Science in society. Yet, insofar as they paid real attention to Bacon's ideas (rather than piously uttering Baconian slogans), the fellows of the early Royal Society ignored the socially directed part of the plan.

CHAPTER 5

1. A complication needs to be noted explicitly. The method for arriving at value-judgments of §7 supposed that participants in an ideal conversation would not suffer from false beliefs. Allowing for any divergence from the truth appears, at first sight, to violate that condition. When the problem is stated, however, the solution is obvious: the range of contexts within which statements must be assessed as to whether they are "true enough" has to include the circumstances of deliberation—accepting the statement must not cause the ideal conversation to reach a different outcome than that achieved under adoption of the exact truth.

2. In proposing a similar perspective in (Kitcher 2001), I began from reflections about Science and made a limited proposal about values. Here, my suggestion is based on a more general conception of values, that sketched in chapter 2 and articulated more fully in (Kitcher 2011a). The earlier route exemplified "science and values" (the common direction for people worried about value-judgments and thus inclined to pose questions about values in science as an afterthought, something forced on them); the present treatment—"values and science"—is less bashful.

3. As that astute logician Charles Dodgson saw, the only complete map would be the terrain itself (Carroll 1976).

4. For a thoughtful exposition of this idea, see (Salmon 1984). I was also beguiled by it (Kitcher 1993). For healthy skepticism, see (Rorty 1982) and (Kitcher 2001).

5. I see the practice of scientific inquiry as framed by the ethical project and as contributing to the further evolution of that project.

6. This is only the schema of a procedure for determining significance, since there are different ways in which the voting could be carried out. Different packages

of investigations could be offered, voters could rank-order the possibilities, they could have a store of points for distribution over options as they saw fit, etc.

7. I owe this anecdote to Gordon Conway, who recounted it in a discussion at the "Living with the Genie" conference, held in 2002 at Columbia University.

8. In terms of the call-to-arms issued by Foucault (see §16), the proposal favors the *integration* rather than the "insurrection" of "knowledges."

9. Thanks to Raine Daston for posing this as a forceful challenge.

10. For this process to achieve the ends proposed, it is obviously necessary that it be set up in ways that avoid replicating the political struggles that need to be overcome. It is entirely reasonable to wonder whether that is possible, and my future remarks about further roles for "deliberative polling" or "citizen juries" will exacerbate the worry. I postpone any attempt to address it until §37.

CHAPTER 6

1. Yet, as we shall see in chapter 8, Feyerabend can be read more sympathetically as calling for a broader variety of perspectives in scientific research.

2. Although you should be very careful to make sure there aren't revealing statistical properties of your announced data sets; two well-known instances of scientific fraud, those of Sir Cyril Burt and of Robert Slutsky, were detected because the means and variances of allegedly distinct data sets were the same.

3. It is worth emphasizing that it is only in special cases (for special types of statements) that one can give a precise sense of proximity to the truth (Kitcher 1993, 120–24).

4. Evelyn Fox Keller has suggested to me that this is too easy a case and that there may well be instances where the propriety of certification is extremely hard to determine. It seems to me important to recognize this possibility. For the present, my principal interest lies in recognizing clear cases of flawed certification and clear cases of unproblematic certification. How large the remainder is, is a task for careful determination (under the aegis of well-ordered science).

5. There are obvious connections here to Rawlsian requirements of publicity (Rawls 1971).

6. The state just described could easily be viewed as the "preparadigm" condition of (Kuhn 1962).

7. Considering the two ideals in tandem allows for the possibility of a more sympathetic reading of Feyerabend—especially of (Feyerabend 1978)—than philosophers of science (including me) have usually provided. I shall not, however, pursue that here.

8. By contrast, philosophers of science who have focused on particular types

of scientific problems, studying them as they arise within current practice, have sometimes arrived at new, widely employable, formal tools. See, for example, the work of Clark Glymour and his associates and of Judea Pearl on causal modeling.

9. Here I recapitulate a point about public reason that has been developed with great sophistication in (Rawls 1996) and (Scanlon 1998).

CHAPTER 7

1. Here I am indebted to Allen Buchanan.

2. Once again, there is a serious concern that the procedure envisaged here will inevitably become mired in the existing political conditions, thus subverting its functioning. As already noted, I postpone discussion of this important worry to §37.

CHAPTER 8

1. Evelyn Fox Keller has charted far more subtle ways in which the presence of women can change a field (and did change primatology). See (Keller 2002). But a relatively straightforward example suffices for my point here.

2. I make this point explicit, since my previous efforts have often been read (possibly understandably) as overemphasizing market forces. Some of the most trenchant criticism of (Kitcher 1990) and (Kitcher 1993) reads me as believing in the magic of markets (Mirowski 2004; Hands 1995). Since I have no such belief, it is worth stating that clearly.

3. This is not, of course, to deny that team playing is a *locally* effective ideal: members of a research group work together, often because they are powerfully motivated to outcompete their rivals! The point is that the members of a scientific specialty do not view themselves as a large "team."

4. The dynamics of the process are those of the Coalition Game, introduced in a very different context in (Kitcher 2011a, sect. 9).

5. It is worth noting that there are some researchers who have attempted to introduce evolutionary ideas into psychology in a much more rigorous and subtle way. See, for example, (Nettle 2011).

6. I am grateful to Meira Levinson for drawing my attention to one of these, the work of James Fishkin on deliberative polling.

CHAPTER 9

1. It should be noted that the techniques of mammalian cloning were not products of contemporary research in molecular medicine. They emerged from the long-established agricultural practice of developing favored stocks (plant cloning, of course, has a long history in agriculture). From the standpoint of ethics and public policy, however, the issues raised by cloning are akin to those in various areas of biomedicine.

2. The "terminator" technology is well explained in (Stewart 2004, 82ff.).

3. According to the Yale survey of attitudes to climate change, 63 percent of US citizens think that climate change is occurring, and 50 percent believe that, if it is occurring, it is caused by human activity. These statistics are sometimes taken as grounds for optimism about public knowledge, but that is to commit a probabilistic fallacy. What the public ought to know is a *conjunction*: global warming is occurring *and* it is caused by human activity. On the basis of answers to the survey questions, it is plain that the class of citizens who accept the conjunction is less than 50 percent of the population—for a conjunction can be no more probable than its least likely conjunct. The percentage of US citizens believing the conjunction depends on the distribution of the belief in the conditional ("if global warming is happening, it is caused by human activity"). If the people who agree with that conditional are predominantly found in the 37 percent who do not think that global warming is occurring, the class of those who accept the conjunction could be significantly less than 50 percent.

REFERENCES

Akerlof, George. 1984. *An Economic Theorist's Book of Tales*. Cambridge: Cambridge University Press.

Ayala, Francisco. 2007. *Darwin's Gift to Science and Religion*. Washington, DC: Joseph Henry Press.

Berlin, Isaiah. 1961. *Two Concepts of Liberty*. Oxford: Clarendon Press.

Bloor, David. 1976. *Knowledge and Social Imagery*. London: Routledge.

Boehm, Christoph. 1999. *Hierarchy in the Forest*. Cambridge, MA: Harvard University Press.

Boyer, Pascal. 2001. *Religion Explained*. New York: Basic Books.

Broad, William, and Nicholas Wade. 1982. *Betrayers of the Truth*. New York: Simon & Schuster.

Brock, William, and Steven Durlauf. 1999. "A Formal Model of Theory Choice in Science." *Economic Theory* 14: 113–30.

Burtt, E. A. 1932. *Metaphysical Foundations of Modern Physical Science*. London: Routledge and Kegan Paul.

Carroll, Lewis. 1976. *Sylvie and Bruno*. New York: Garland.

Cartwright, Nancy. 1999. *The Dappled World*. Cambridge: Cambridge University Press.

Cheney, Dorothy, and Robert Seyfarth. 1990. *How Monkeys See the World*. University of Chicago Press.

Collins, Harry. 1985. *Changing Order*. London: Sage.

Cosmides, Leda, and John Tooby. 1992. "Cognitive Adaptations for Social Exchange." In *The Adapted Mind*, edited by Jerome Barkow, Leda Cosmides, and John Tooby. New York: Oxford University Press.

Coyne, Jerry. 2009. *Why Evolution Is True*. New York: Viking.

Craig, Edward. 1990. *Knowledge and the State of Nature*. Oxford: Oxford University Press.

Dahl, Robert. 1963. *A Preface to Democratic Theory*. University of Chicago Press.

———. 1970. *After the Revolution?* New Haven: Yale University Press.

Dawkins, Richard. 1998. *Unweaving the Rainbow*. Boston: Houghton Mifflin.

———. 2006. *The God Delusion*. New York: Houghton Mifflin.

———. 2009. *The Greatest Show on Earth*. New York: Free Press.

Dennett, Daniel. 2006. *Breaking the Spell*. New York: Penguin.

Derrida, Jacques. 1976. *Of Grammatology*. Baltimore: Johns Hopkins.

De Waal, Frans. 1996. *Good-Natured*. Cambridge, MA: Harvard University Press.
Dewey, John. 1934. *A Common Faith*. New Haven: Yale University Press.
———. 1958. *Problems of Man*. Paterson, NJ: Littlefield Adams.
———. 1985. *The Public and Its Problems*. Athens, OH: Swallow Press.
———. 1997. *Democracy and Education*. New York: Free Press.
Douglas, Heather. 2009. *Science, Policy, and the Value-Free Ideal*. University of Pittsburgh Press.
Dupre, John. 1993. *The Disorder of Things*. Cambridge, MA: Harvard University Press.
Dworkin, Ronald. 1993. *Life's Dominion*. New York: Knopf.
Earman, John. 1992. *Bayes or Bust?* Cambridge, MA: MIT Press.
Eldredge, Niles. 1982. *The Monkey Business*. New York: Washington Square Press.
Elgin, Catherine. 2004. "True Enough." *Philosophical Issues* 14: 113–31.
Engler, Robert et al. 1988. "Misrepresentation and Responsibility in Medical Research." *New England Journal of Medicine* 317: 1383–89.
Epstein, Stephen. 1996. *Impure Science*. University of California Press.
Estlund, David. 2008. *Democratic Authority*. Princeton University Press.
Feyerabend, Paul. 1975. *Against Method*. London: Verso.
———. 1978. *Science in a Free Society*. London: New Left Books.
Fisher, R. A. 1936. "Has Mendel's Work Been Rediscovered?" *Annals of Science* 1: 115–37.
Fishkin, James. 2009. *When the People Speak*. New York: Oxford University Press.
Flory, James, and Philip Kitcher. 2004. "Global Health and the Scientific Research Agenda." *Philosophy and Public Affairs* 32: 36–65.
Foucault, Michel. 1980. *Power/Knowledge*. New York: Pantheon.
Galton, Francis. 1875. *English Men of Science*. New York: Appleton.
Goldman, Alvin. 1986. *Epistemology and Cognition*. Cambridge, MA: Harvard University Press.
———. 1999. *Knowledge in a Social World*. New York: Oxford University Press.
Goodall, Jane. 1988. *The Chimpanzees of Gombe*. Cambridge, MA: Harvard University Press.
Goodin, Robert. 2003. *Reflective Democracy*. Oxford: Oxford University Press.
Gould, Stephen Jay. 1977. *Ever Since Darwin*. New York: Norton.
———. 1981. *The Mismeasure of Man*. New York: Norton.
Gross, Paul, and Norman Levitt. 1994. *Higher Superstition*. Baltimore: Johns Hopkins.
Hands, D. Wade. 1995. "Social Epistemology Meets the Invisible Hand." *Dialogue* 34: 605–22.
Hansen, James. 2009. *Storms of My Grandchildren*. New York: Bloomsbury.
Haraway, Donna. 1989. *Primate Visions*. New York: Routledge.

Haufe, Christopher. Forthcoming. *The Sound of Science*. Ms.
Hempel, C. G. 1966. *Philosophy of Natural Science*. Englewood Cliffs: Prentice-Hall.
Herman, Edward S., and Noam Chomsky. 1988. *Manufactured Consent*. New York: Pantheon.
Hitchens, Christopher. 2007. *God Is Not Great*. New York: Twelve.
Holtzman, N. A. 1989. *Proceed with Caution*. Baltimore: Johns Hopkins.
Horkheimer, Max, and Theodor Adorno. 1978. *Dialectic of Enlightenment*. New York: Seabury.
James, William. 1984. *Writings 1902–1910*. Library of America.
Jasanoff, Sheila. 2005. *Designs on Nature*. Princeton University Press.
Jefferson Project Report. Available at http://www.jefferson-center.org/index.asp?Type=B_LIST&SEC=%7BFE67AE78-C3E6-4A12-94E1-A2CF75FB82B9%7D.
Jeffrey, Richard. 1956. "Valuation and Acceptance of Scientific Hypotheses." *Philosophy of Science* 22: 237–46.
Judson, Horace Freeland. 1979. *The Eighth Day of Creation*. New York: Simon & Schuster.
———. 2004. *The Great Betrayal*. Orlando: Harcourt.
Kamin, Leon J. 1974. *The Science and Politics of I.Q.* Potomac, MD: Erlbaum.
Keller, Evelyn Fox. 2009. *The Mirage of a Space between Nature and Nurture*. Durham: Duke University Press.
———. 2002. "Women, Gender, and Science: Some Parallels between Primatology and Developmental Biology." Chapter 8 in *Primate Encounters: Models of Science, Gender, and Society*, edited by Shirley C. Strum and Linda Marie Fedigan. Chicago: University of Chicago Press.
———. Forthcoming. "Climate Science, Truth, and Democracy." Ms.
Kevles, Daniel. 1978. *The Physicists*. New York: Knopf.
Kitcher, Philip. 1982. *Abusing Science*. Cambridge, MA: MIT Press.
———. 1984."1953 and All That. A Tale of Two Sciences." *Philosophical Review* 93: 335–73.
———. 1985. *Vaulting Ambition: Sociobiology and the Quest for Human Nature*. Cambridge, MA: MIT Press.
———. 1990. "The Division of Cognitive Labor." *Journal of Philosophy* 87: 5–22.
———. 1993. *The Advancement of Science*. New York: Oxford University Press.
———. 1996. *The Lives to Come*. New York: Simon & Schuster.
———. 1999. "The Hegemony of Molecular Biology." *Biology and Philosophy* 14: 195–210.
———. 2001. *Science, Truth, and Democracy*. New York: Oxford University Press.
———. 2005. "The Hall of Mirrors." *Proceedings and Addresses of the American Philosophical Association* 79: 2, 67–84.

———. 2007. *Living with Darwin*. New York: Oxford University Press.
———. 2009. "Carnap and the Caterpillar." *Philosophical Topics* 36 (1).
———. 2010. "Varieties of Altruism." *Economics and Philosophy* 26 (2).
———. 2011a. *The Ethical Project*. Cambridge, MA: Harvard University Press.
———. 2011b. "Challenges for Secularism." In *The Joy of Secularism*, edited by George Levine. Princeton University Press.
Knauft, Bruce. 1991. "Violence and Sociality in Human Evolution." *Current Anthropology* 32: 391–428.
Koertge, Noretta, ed. 1998. *A House Built on Sand*. New York: Oxford University Press.
Kuhn, Thomas S. 1962. *The Structure of Scientific Revolutions*. University of Chicago Press.
Latour, Bruno. 1987. *Science in Action*. Cambridge, MA: Harvard University Press.
Laudan, Larry. 1984. *Science and Values*. University of California Press.
Lee, Richard. 1979. *The !Kung San*. Cambridge University Press.
Leuschner, Anna. 2011. *Wissenschaftliche Glaubwürdigkeit*. PhD dissertation, Universität Bielefeld.
Levi, Isaac. 1960. "Must the Scientist Make Value Judgments?" *Journal of Philosophy* 57: 345–57.
———. 2011. "Knowledge as True Belief." In *Belief Revision Meets Philosophy of Science*, edited by Erik J. Olsson and Sebastian Enqvist. New York: Springer.
Lewontin, Richard C. 2002. "The Politics of Science." *New York Review of Books*, May 9.
Longino, Helen. 1990. *Science as Social Knowledge*. Princeton University Press.
Lyotard, Jean-François. 1984. *The Post-Modern Condition*. Minneapolis: University of Minnesota Press.
MacKinnon, Catherine. 1989. *Toward a Feminist Theory of the State*. Cambridge, MA: Harvard University Press.
Merton, Robert. 1968. *Social Theory and Social Structure*. New York: Free Press.
Mill, John Stuart. 1970. *Collected Works of John Stuart Mill*. Volumes 2–3. Toronto: University of Toronto Press.
———. 1998. *On Liberty and Other Essays*. New York: Oxford University Press [World's Classics].
Miller, Kenneth. 1999. *Finding Darwin's God*. New York: Cliff Street Books.
Milton, John. 1963. *Milton's Prose*. London: Oxford University Press [World's Classics].
Mirowski, Philip. 2004. "The Economic Consequences of Philip Kitcher." In *The Effortless Economy of Science*, by P. Mirowski. Durham: Duke University Press.
Nelkin, Dorothy, and Laurence Tancredi. 1994. *Dangerous Diagnostics*. University of Chicago Press.

Nettle, Daniel. 2011. "Flexibility in Reproductive Timing in Humans." *Proceedings of the Royal Society B*. In press.

Norris, Stephen, and Linda Phillips. 2003. "How Literacy in Its Fundamental Sense Is Central to Scientific Literacy." *Science Education* 87: 224–40.

Numbers, Ronald. 1992. *The Creationists*. New York: Knopf.

Olby, Robert. 1974. *The Path to the Double Helix*. Seattle: University of Washington Press.

Oreskes, Naomi, and Erik Conway. 2010. *Merchants of Doubt*. New York: Bloomsbury.

Pettit, Philip. 1997. *Republicanism*. Oxford: Oxford University Press.

Plato. 1992. *Republic*. Indianapolis: Hackett.

Rawls, John. 1971. *A Theory of Justice*. Cambridge, MA: Harvard University Press.

———. 1996. *Political Liberalism*. New York: Columbia University Press.

Reichenbach, Hans. 1949. *Experience and Prediction*. University of Chicago Press.

Reiss, Julian, and Philip Kitcher. 2009. "Biomedical Research, Neglected Diseases, and Well-Ordered Science." *Theoria* 66: 263–82.

Richardson, Henry. 2002. *Democratic Autonomy*. New York: Oxford University Press.

Rorty, Richard. 1982. *Consequences of Pragmatism*. Minneapolis: University of Minnesota Press.

Rudner, Richard. 1953. "The Scientist *qua* Scientist Makes Value Judgments." *Philosophy of Science* 20: 1–6.

Rudwick, M. J. S. 1985. *The Great Devonian Controversy*. University of Chicago Press.

Ruse, Michael. 1982. *Darwinism Defended*. Reading, MA: Addison-Wesley.

———. 2005. *The Evolution-Creation Struggle*. Cambridge, MA: Harvard University Press.

Salmon, Wesley. 1984. *Scientific Explanation and the Causal Structure of the World*. Princeton University Press.

Sayre, Anne. 1975. *Rosalind Franklin and DNA*. New York: Norton.

Scanlon, Thomas M. 1998. *What We Owe to Each Other*. Cambridge, MA: Harvard University Press.

Schneider, Stephen. 2009. *Science as a Contact Sport*. Washington, DC: National Geographic.

Schumpeter, Joseph. 1947. *Capitalism, Socialism, and Democracy*. New York: Harper.

Shapin, Steven, and Simon Schaffer. 1985. *Leviathan and the Air-Pump*. Princeton University Press.

Shapiro, Ian. 2003. *The Moral Foundations of Politics*. New Haven: Yale University Press.

Shostak, Marjorie. 1981. *Nisa*. Cambridge, MA: Harvard University Press.

Sokal, Alan, and Jean Bricmont. 1998. *Fashionable Nonsense*. New York: Picador.

Stewart, C. Neal. 2004. *Genetically Modified Planet*. New York: Oxford University Press.

Strevens, Michael. 2003. "The Role of the Priority Rule in Science." *Journal of Philosophy* 100: 55–79.

Vickers, A. Leah, and Philip Kitcher. 2002. "Pop Sociobiology Reborn: The Evolutionary Psychology of Sex and Violence." In *Evolution, Gender, and Rape*, edited by Cheryl Travis. Cambridge, MA: MIT Press.

Watson, James D. 1968. *The Double Helix*. New York: Atheneum.

Westermarck, Edward. 1924. *Origin and Development of the Moral Ideas*. 2 Volumes. London: Macmillan.

Wilholt, Torsten. 2009. "Bias and Values in Scientific Research." *Studies in History and Philosophy of Science* 40: 92–101.

Wylie, Alison. 2000. "Rethinking Unity as a Working Hypothesis for Philosophy of Science." *Philosophical Perspectives* 7: 293–317.

INDEX

Italicized entries correspond to the original definition of a term or phrase, or to subsequent extensions of its usage.